The Quintessential Question

I dedicate this book to grandparents, relatives, friends, and teachers.

K.R.

ISBN-10: 1548587435

ISBN-13: 978-1548587437

Printed in the United States

Font Set in Maiandra GD, Arabic Typesetting, Candara, Calibri, Ebrima

Summary: This is a book of geographic questions and vital information designed to prepare students in grades 4-8 competing in the School, State, and National Geographic Bees, Junior/Senior NSF Geography Bees, U.S. Geography Olympiad, and International Geography Bee.

Design and Text by Keshav Ramesh

Cover Illustration by Keshav Ramesh

THE QUINTESSENTIAL QUESTIONNAIRE TO THE GEOGRAPHY BEE

by Keshav Ramesh

The Quintessential Questionnaire to the Geography Bee

The Geography Bee Ultimate Preparation Guide and A Competitor's Compendium to the Geography Bee:

This young genius has come up with a great guide! His questions follow the various topics that come up on the bee and simulate the types of questions that come.

A must-buy for the Geography Bee contestant!

- **Karan Menon,** 2015 NGB Champion, 2015 U.S. Geography Olympiad JV Champion, and 2017 National U.S. International Geography Bee Third Place Winner

The Geography Bee Ultimate Preparation Guide:

These questions really helped me in my geography endeavors. Keshav is a great author who made many great questions. The content is very well organized and has many different tips and questions. The questions are easy to study from and can help you become smarter in no time!

- **Rohit Gunda,** 2016 Connecticut State Geographic Bee Third Place Winner

The Geography Bee Ultimate Preparation Guide and A Competitor's Compendium to the Geography Bee:

Your books are so useful! I got 3rd at the NJ state bee using just them and travel videos and an atlas. I encourage you to check out books written by Keshav Ramesh, which each has specific sections designed for different levels of the bee. I really appreciated Keshav's book, and it taught me a lot of things. It's a very good book.

-**Ken Mitchell,** 2017 New Jersey State Geographic Bee Third Place Winner

Table of Contents

Tips, Tricks, and How to Prepare for the Geography Bee

What materials should I have when studying geography?

When studying for the geography bee, I would recommend keeping an atlas, detailed political map, and detailed physical map near you. These three are the most vital sources you need to help you achieve success in the National Geographic Bee.

From looking at a detailed political map, you can find cities and countries from around the world. Territories, islands, and dependencies will be included. You can find oceans, seas, lakes, straits, gulfs, bays, and even rivers. These will be emphasized in a physical map.

There are also smaller versions of political maps – for example, ones showing just South Asia or Central America.

Archipelagos are scattered across the world, like Indonesia and Japan. A political map will show you the countries in different colors, and they will be labeled.

Political maps will also show you the continents, and the names of countries will be printed much bigger than the cities.

A physical map will also help as well, to identify the landforms (which will be labeled) as well as biomes. Ocean ridges and seamounts can be displayed on physical maps but not necessarily on all. Different colors will be expressed on the map, as well as ridges to show mountains, or a plain, yellowish color spreading across a certain area to signify a desert. Also shown on physical maps are bodies of water, plateaus, geographical regions, basins, and other major parts of the Earth's topography.

Detailed physical maps, or ones that show a close-up of a region or country in the world, also have reservoirs, smaller bays/gulfs, highlands, hills, passes, and much more.

Atlases are great for researching thousands of facts just by looking at maps and diagrams, and reading about them. Countries are always featured in atlases, and you'll always find a few chapters about the physical geography of the world.

You will find many thematic maps in atlases, such as those depicting religions, language families, economy, water supply, major agricultural commodities, and more.

I would recommend buying atlases sold by National Geographic, as they, in my opinion, provide the best information.

The National Geographic Atlas of the World, Tenth Edition is good for participants serious about winning the state bees and doing well in the national bees. No atlas like this in the world has a lot of very detailed information. Although it is expensive, it is worth the price. State winners have been awarded this book in the past.

You should get an almanac, like the *National Geographic Kids Almanac 2018* or another good one recently published from National Geographic with interesting (and sometimes weird!) geography facts.

However, this book gives you a broader outlook on geography, so it is a good reference for your school and sometimes even your state bees.

The *National Geographic Magazine* is especially important, as questions regarding facts in the magazine have been asked.

If you are participating in the 2018 National Geographic Bee, learn the geographical facts from the magazine from its six issues before May (November 2017-April 2018). This applies to 2019 and beyond as well.

You should also get a book about the physical geography of the world, as this is vital to the physical geography section of the National Geographic Bee. The *National Geographic Desk Reference* is probably the best option for you when looking at physical geography.

How difficult are the questions in this book?

The difficulty of these questions range from the preliminary rounds of the State Competition to the final rounds of the National Bee. You should use these questions to prepare to do well in and/or win your state and national bees. However, year after year, a growth in difficulty has appeared in the school bee rounds.

The questions increase in difficulty as you finish more and more, before rivaling questions that can be found in the national championship rounds.

However, in the chapters concerning cultural geography, economic geography, etc., the questions will be more or less of the same difficulty – there is a possibility of encountering, for example, a very easy physical geography question and moving on to the next one only to find that it is quite difficult. This is actually representative of what happens in the bee – there is a lot of luck involved! Don't be disheartened if you answer a very difficult question incorrectly and the next person gets a question asking for a country's capital.

Sometimes the questions will be easier than the preceding ones in this book, but it is good to be prepared for whatever question is given to you during any stage of the competition.

In other words, only some of the questions in this book will be in order of difficulty, and most of the questions will be in no specific order.

How many questions are in this book?

The Quintessential Questionnaire to the Geography Bee has **over 2,150 questions to help you in all levels of the National Geographic Bee!** The questions are separated by continent and by each topic in geography. There are also subtopics in some of the geographical topic chapters (Physical Geography, Cultural Geography, Economic Geography, etc.).

What is the daily amount of time I should study?

If you want to win the school bee, studying for 30 minutes to 1 hour a day is enough (unless you are in a competitive school for the geography bee for those of you in New Jersey, Florida, California, Washington, Michigan, Virginia, Maryland, or Texas). For the state level bee, I would recommend studying 2 to 3 hours a day.

If you are at the level of the National Bee Preliminaries, 3 to 4 hours is what I would recommend. Those aspiring to achieve a place in the top ten should definitely amount to more than 4 hours, if not 5-7.

This competition, like others, is challenging. Hours of dedication to geography is vital to your success in the state and national bees.

Is this book only for competitors in the National Geographic Bee?

This book is not only for people competing in the National Geographic Bee, but for other competitions as well.

If you are participating in the finals of the **North South Foundation's Junior/Senior Geography Bees,** this is a good book for you. There is a separate chapter for Indian Geography.

This book contains questions not only written for the National Geographic Bee, but also for the **United States Geography Olympiad (USGO)**. The USGO starts regionally as a National Qualifying Exam and any competitor must score above the national median score and/or in the top half at the regional level.

The **National USGO Championships** are held along with the **National History Bee and Bowl,** and the U.S. International Geography Bee in Arlington, Virginia annually in April. This

Olympiad selects the top four individuals in the Varsity Division to represent the United States at the **International Geography Olympiad (iGeo)**, held in a different city around the world annually.

This book has also been written for students preparing for the **International Geography Bee (IGB)**, an intense geography competition created in 2017 with a regional National Qualifying Exam and fast quiz bowl rounds at the national level.

The U.S. national championships of the International Geography Bee is held in conjunction with the USGO and the **National History Bee and Bowl** annually in Arlington, Virginia in April. The top four individuals in the JV Division represent the United States region at the **Junior Varsity IGB World Championships.** The top four individuals in the Varsity division represent the United States region at the **Varsity IGB World Championships.**

As a list, this book is a great resource for these competitions:

- National Geographic Bee (NGB)
- NSF Junior/Senior Geography Bees (NSF)
- United States Geography Olympiad (USGO)

- International Geography Bee (IGB)

What should I study for the state and national competitions of the National Geographic Bee?

For the state bee, knowledge of borders, locations, and the United States is important. Be sure to know your national parks, national forests, and national monuments. You should be well versed in the major cities of each state and country.

Be sure to follow National Geographic's Instagram and Twitter for more geographic information. **In the 2016 State Geographic Bees, questions regarding photos from National Geographic's Instagram account were posed to the competitors in the final rounds.**

If you reach the finals of the state bee, your atlas should be your greatest ally. You should have a complete mastery or near mastery of thousands and thousands of locations – however, this should not be achieved through memorization, but through constant review. It may sound difficult, but it gets much easier after you practice and practice, at least three hours a day.

Although knowing thousands of locations sounds like a lot, you probably know the name and location of every single country, some major cities, major rivers, major lakes, major mountain ranges, notable islands, deserts, oceans, and seas if you are an experienced geography bee participant. This by itself is a lot of information.

Be thorough with your atlases, and use the National Geographic Atlas of the World, Tenth Edition as much as possible.

Create questions, research facts, connect locations, and take notes. Keep rereading what you have read and written down/typed and try to produce questions from the facts. Integrate the facts and questions as much as possible so that you can remember them.

You should be thorough with tourist attractions and landmarks, as well as river confluences, port cities, UNESCO World Heritage Sites, major exports, current events, physical geography terms, and geographic extremes (such as Kanyakumari, Cape Byron, and Ushuaia).

What do I need to know about each country?

The Quintessential Questionnaire to the Geography Bee

Here's a list of what you need to know about each country to help you prepare for the National Geographic Bee. Remember, some of the things on this list may not apply to certain countries:

Basics:
- Location (Continent)
- Location (Region, e.g. South Asia)
- Capital
- Major Cities (At least 10+ for countries with populations over 70,000,000)
- Population (Approximate)
- Official Country Name

Physical:
- Highest and Lowest Points
- Mountain Ranges, Peaks, and Volcanoes
- Rivers, Deltas, River Mouths/Sources
- Lakes, Reservoirs, and Dams
- Bordering Seas
- Gulfs and Bays
- Straits, Sounds and Inlets
- Plateaus
- Peninsulas, Capes, and Points
- Plains and Basins
- Wetlands, Swamps, and Marshes
- Deserts, Ergs, and Dunes

- Valleys and River Valleys
- Grasslands and Prairies
- Waterfalls
- Canyons and Gorges
- Islands and Archipelagoes
- Isthmuses and Spits
- Canals
- Physical Regions
- Buttes and Mesas
- Major Glaciers and Fjords
- Lagoons and Reefs
- Continental Divides, Earth Physical Structure, Layers of the Earth

Political
- Bordering Countries
- Administrative Divisions (States, Provinces, Federal Districts, Counties)
- Territories, Dependencies, and Occupied Atolls
- Current Leader(s) (President, Prime Minister, King/Queen, Chairman, Prince, etc.)
- Government Structure/Important and Influential Laws and Type of Rule (Republic, Democracy, Monarchy, etc.)
- Disputed Countries and Regions
- Politically Established Regions (Northeast Africa, Central America, etc.)

- National and Global Organizations

Cultural
- Religions
- Languages
- Festivals, Holidays, and Traditions
- Foods, Art, Music, and Architecture
- Cultural Items/Objects and Symbols

Environmental
- Oceans
- Conservation and Biodiversity
- Biomes and Habitats
- Plants and Animals
- Global Warming/Climate Change
- Environmental Hot Spots
- Natural Disasters

Economic
- Currencies
- Trade, Exports, and Imports
- Production
- Agricultural Products and Natural Resources
- Port Cities, Seaports, and International Airports

Historical:
- Kingdoms and Empires

- Wars
- Former Countries, Historical Colonies, Territories, Past Leaders, Independence

Landmarks
- National Parks/Preserves, Forests, and Monuments
- National Historic Sites and Historical Parks (United States)
- UNESCO World Heritage Sites
- Famous Castles/Ruins, Museums, and Zoos
- Space Centers, Observatories, and National Laboratories

Current:
- Global, Nationwide, and Environmental Issues
- Archaeological Discoveries
- Climate Change and Global Warming
- International Relations
- Major Worldwide Sporting Events (FIFA World Cup Russia 2018, Tokyo 2020 Summer Olympics)

What are some good websites I can use to prepare?

The website I'd recommend to help you prepare for the competition is www.nationalgeographic.com/geobee. It has great tools to help you study, and also a page where you can play a game called the GeoBee Challenge, where National Geographic gives you 10 new questions every day to prepare with.

National Geographic has also partnered with Kahoot to create geography bee prep games. Below is a list of all websites you should probably use to enhance your knowledge.

Sites Dedicated Exclusively to Geography and the Geography Bee:

www.nationalgeographic.org/geobee

www.geobeeworld.blogspot.com

www.geography.about.com

http://lizardpoint.com/geography/

http://www.sporcle.com/games/category/geography/all

http://nationalgeographic.org/bee/study/play-kahoot/

What are some other online resources I can use?

Quizlet is an online quizzing program that you can use to prepare for the geography bee. I would also use Socrative, where you can make multiple choice questions with multiple answers (if you choose). There are hundreds, maybe thousands of geography quizzes on Sporcle.

The world's largest geography bee community is on Google+. Look up **GeoBee City** on Google+, and you will find it. GBC is a geography bee community where students from all over the country work on geography-related projects, mock bees, quizzes, maps, and use resources to study. GBC was founded by Karan Menon, the 2015 National Geographic Bee Champion, in 2014 and the community has over 200 members.

In addition to studying for the National Geographic Bee with fellow students, we prepare for the **United States Geography Olympiad** and the **International Geography Bee**, and some of us who are eligible prepare for the **North South Foundation (NSF) Junior and Senior Geography Bees.**

Any questions or concerns?

Any questions you'd like me to add? Still need help with geography? Any errors that need to be fixed? Contact me at keshav.ramesh@gmail.com!

Is there anything else I should know when preparing for the geography bee?

There is!

Let us set the scenario that you were participating in the 2012 State Geographic Bee in fourth grade, up to the 2016 State Geographic Bee in eighth grade. This way, National Geographic has a near-perfect way of guaranteeing that you wouldn't have known some of the questions asked in the 2011 State Geographic Bee finals as you would've been in third grade and most likely not watched the final rounds.

Let us look at a different scenario. Let's say you were participating in the 2013 State Geographic Bee in fourth grade, up to the 2017 State Geographic Bee in eighth grade. Again, wouldn't National Geographic have a near-perfect way of guaranteeing that you wouldn't have known some of the questions asked in the 2012 State Geographic Bee? So instead of creating a set of new final round questions, they could just reuse them.

Certainly, the National Geographic Society would have to create new questions for the state finals – but not all of the questions would necessarily have to be created as of that year.

So for those of you participating in the 2018, 2019, or 2020 State Geographic Bee, I would suggest looking at these videos below now or sometime later, and make sure that you record the questions, test yourself on them, and maybe find some other patterns between the questions of different years:

2013 for 2018:

http://ct-n.com/ctnplayer.asp?odID=8880

2014 for 2019:

https://www.youtube.com/watch?v=rJX_YSSOVU8

2015 for 2020:

https://www.youtube.com/watch?v=Tu3xJLqJIFA

2016 for 2021:

https://www.youtube.com/watch?v=6FuT9nAl9dY

2017 for 2022:

http://ct-n.com/ctnplayer.asp?odID=13916

Should I use National Geographic's Instagram account to help me study?

Using National Geographic's Instagram account to learn geography is a great idea. (you don't need an Instagram account to do this!).

Many questions are actually based off of pictures National Geographic posts of oceans, animal and plant species, conservation projects, geography, science, and nature. Check out these links!:

https://www.instagram.com/natgeo/

https://www.instagram.com/natgeotravel/

https://www.instagram.com/natgeoyourshot/

https://www.instagram.com/natgeowild/

Good luck and happy studying for the National Geographic Bee!

United States/State

Savvy

1. What is the largest city by population in the panhandle of Connecticut?
    Stamford

2. Reno is a major city in what U.S. state?
    Nevada

3. Mystic Seaport is located in the southern part of what New England state?
    Connecticut

4. What state is famous for its Cajun culture and borders the Gulf of Mexico?
    Louisiana

5. Name the only national park that in South Carolina.
    Congaree National Park

6. Tampa and St. Petersburg are major cities in what state?
    Florida

7. Padre Island belongs to what state with a panhandle?

Texas

8. The Aleutian Islands are an archipelago belonging to what Pacific state?
Alaska

9. The Kittatinny Mountains form the Wallpack Valley in what Mid-Atlantic state?
New Jersey

10. You can ride a train at the Jelly Belly warehouse near the city of Kenosha in which state?
Wisconsin

11. The Natural Bridge Caverns, the largest known commercial caverns in Texas, are near what major city that is in the southwestern part of the Texas Triangle?
San Antonio

12. You can create your own candy bar at Hershey's Chocolate World near Harrisburg in which state?
Pennsylvania

13. Stonehenge II, a model of the original Stonehenge in the United Kingdom, is located in what state where Cadillac Ranch can be found?
Texas

14. Lake Borgne is a lagoon of the Gulf of Mexico in what U.S. state?

Louisiana

15. Buffalo wings take their name from a city located on Lake Erie in which state?
New York

16. The Mahomet Aquifer is the most important aquifer in the eastern part of what state bordering Lake Michigan to the north?
Illinois

17. You can create your own custom PEZ candy dispenser near New Haven in which New England state?
Connecticut

18. Mille Lacs Lake is the second largest inland lake in what U.S. state whose largest is Red Lake?
Minnesota

19. Oreo cookies are mainly produced in Richmond in which eastern state?
Virginia

20. The World's Tallest Thermometer is located near Death Valley and the Cronese Mountains in the city of Baker in what state?
California

21. The Shivwits Plateau is located near the Hurricane Cliffs in what state's Mohave County?

Arizona

22. Diamond Peak is the highest point in the Lemhi
    Mountain Range in what U.S. state?
    Idaho

23. The Buttermilk Channel in southeastern New York
    separates what island from Brooklyn?
    Governors Island

24. The Union Watersphere is near Newark Liberty
    International Airport in what state?
    New Jersey

25. Cape Krusenstern National Monument is in
    northwestern Alaska along what sea?
    Chukchi Sea

26. Doritos were invented in the 1960s at Disneyland in the
    Santa Ana Valley in which state?
    California

27. The Palo Duro Canyon, part of the Caprock Escarpment,
    is located near Amarillo in what state?
    Texas

28. You can find a high concentration of hoodoos in Goblin
    Valley State Park in the San Rafael Swell of what state?
    Utah

29. In 1989, Hurricane Hugo tore through the southeastern United States, nearly destroying Francis Marion National Forest in what state bordering North Carolina?
    South Carolina

30. LaBarge Rock is in the shape of a column and rises above the Missouri River in Lewis and Clark National Forest in what state?
    Montana

31. Omaha, the seat of Nebraska's Douglas County, is located on what river?
    Missouri River

32. The Dr. Pepper Museum is located in Waco near the Brazos River in which state?
    Texas

33. A Jell-O plant is located in Mason City in which state that borders the Mississippi River?
    Iowa

34. The city of Taos is located east of the Rio Grande Gorge and near the Taos Plateau Volcanic Field in what state?
    New Mexico

35. Name the deepest lake in Maine, home to Frye Island and bordering a state park of the same name.
    Sebago Lake

36. The World's Largest Buffalo Monument is a sculpture located in Jamestown, in what state bordering Montana?
North Dakota

37. Part of the Pryor Mountains are located in Custer National Forest in what state home to Medicine Rocks State Park?
Montana

38. Itasca Park in Minnesota is where you can find the headwaters of what major river?
Mississippi River

39. The St. Francis River can be found flowing near Taum Sauk Mountain, the highest natural point in what state?
Missouri

40. The Tooth of Time is a landform in the Sangre de Cristo Mountains in what state?
New Mexico

41. The Oxnard Plain borders the Topatopa Mountains to the north and is located near what cape in a state park of the same name?
Point Mugu

42. Potato chips are produced by the Utz factory in Hanover in which state?
Pennsylvania

43. Davis Dam forms what reservoir in the Cottonwood Valley between Arizona and Nevada?
   Lake Mohave

44. The Ben & Jerry's ice cream factory is located east of Lake Champlain in which state?
   Vermont

45. Lake Hartwell is situated on the border between Georgia and what other state?
   South Carolina

46. Thor's Well is an odd sinkhole formation on Cape Perpetua in what northwestern state?
   Oregon

47. The Pecos River originates in what state's part of the Sangre de Cristo Mountains?
   New Mexico

48. Passamaquoddy Bay, located at the mouth of the St. Croix River, is an inlet of what bay that borders Maine and has the highest tidal range in the world?
   Bay of Fundy

49. Seitz Canyon can be found in the Ruby Mountains near Seitz Lake in Elko County in what state?
   Nevada

50. What state, known as the Centennial State, is home to the San Luis Valley and Glenwood Canyon?
Colorado

51. Chauncey Peak is part of the Metacomet Ridge in what New England state?
Connecticut

52. Kawai Nui Marsh, the largest wetland area in Hawaii, is on what island where the Lanikai Beach can be found?
Oahu

53. The Mojave Phone Booth, a lone telephone booth that was removed in the year 2000, was located in Mojave National Preserve in what state?
California

54. Carhenge is a replica of Stonehenge in the United Kingdom located in the High Plains Region, near the city of Alliance in what state?
Nebraska

55. Lucy the Elephant is a 65-foot tall structure near Atlantic City in what state?
New Jersey

56. Walpole Island is located in Canada's Ontario Province in what lake bordering Michigan to the west?
Lake St. Clair

57. The Chamizal National Memorial is situated on the border between the United States and Mexico in what Texan city?
El Paso

58. The Cabazon Dinosaurs are two large dinosaur sculptures near the San Gorgonio Pass in Cabazon in what state?
California

59. Sowbelly Canyon is part of the Pine Ridge Region in what state whose panhandle is home to the Courthouse and Jail Rocks?
Nebraska

60. You can watch underwater performances by "mermaids" at the Weeki Wachee Springs in what state bordering Alabama to the northwest?
Florida

61. Tallulah Falls Lake is located above Tallulah Gorge in what state whose highest point is Brasstown Bald?
Georgia

62. Czechland Lake Recreation Area and Conestoga Lake are bodies of water in what state home to the Nine Mile Prairie?
Nebraska

63. San Timoteo Canyon is northeast of the Redlands in the San Jacinto Mountains of what state?
California

64. The Menominee River enters Green Bay between Michigan and what other state?
Wisconsin

65. The Balcones Fault System, thought to be related to the formation of the Ouachita Mountains, is located in what state?
Texas

66. Currituck Sound, an inlet of the Atlantic Ocean, is located north of Albemarle Sound and borders North Carolina and what other state?
Virginia

67. Pahuk Hill is situated on the Platte River in what state whose Robidoux Pass is located in the Wildcat Hills?
Nebraska

68. The Meramec Caverns are located within the Ozark Mountains in what state?
Missouri

69. Cavanal Hill is known as the world's tallest hill and is located in what state home to the Antelope Hills?
Oklahoma

70. McKittrick Canyon is situated in what Texan mountain range bordering the Delaware Mountains to the south?
Guadalupe Mountains

71. Smith Falls is located next to the Niobrara National Scenic River in what state?
Nebraska

72. The Upper and Lower Peninsulas in Michigan are connected by what bridge that spans the Straits of Mackinac?
Mackinac Bridge

73. Santa Rosa Sound borders the Fairpoint Peninsula in what state?
Florida

74. The Tahquamenon Falls are a set of waterfalls near Lake Superior in what state?
Michigan

75. Great Captain Island is situated off the coast of Greenwich and contains the southernmost point of land in New England. This island belongs to what state?
Connecticut

76. Mount Shasta is in the Siskiyou County of what state?
California

77. Lucy the Elephant is the oldest roadside tourist attraction in the United States on Absecon Island in what state?
New Jersey

78. Lake Winnipesaukee includes Paugus Bay in the Lakes Region of what state?
New Hampshire

79. Zwaanendael Colony, a Dutch colonial settlement, was built in 1631 in what state known as the "Blue Hen State"?
Delaware

80. The city of Russellville is located on Lake Dardanelle in what landlocked state?
Arkansas

81. Brasstown Bald is the highest peak in what state that is the largest state east of the Mississippi River?
Georgia

82. The Arikaree River, a tributary of the Republican River, has its source in Elbert County in what state?
Colorado

83. The Bayou Corne Sinkhole is located in the Assumption Parish of what state?
Louisiana

84. What state is known as "America's Dairyland", and has Timms Hill as its highest point?
Wisconsin

85. Waugoshance Point borders Sturgeon Bay on the northwestern coast of what peninsula in Michigan?
Lower Peninsula

86. What Hawaiian island is famously regarded as "The Gathering Place" and is home to the Hawaii Cryptologic Center and Kualoa Regional Park?
Oahu

87. Mono Lake is a saline soda lake in the Mono Basin of what state home to the Santa Cruz Mountains?
California

88. What state preserves Puebloan structures at Aztec Ruins National Monument near the city of Aztec?
New Mexico

89. Butterfield Canyon is located in the Oquirrh Mountains of what state home to Delicate Arch and Kings Peak?
Utah

90. Turner Falls is located in the Arbuckle Mountains, an ancient mountain range in what state?
Oklahoma

91. The Monument Rocks, also known as the Chalk Pyramids, are located in what state home to Big Basin Prairie Reserve?
Kansas

92. Attu Island, the largest of the Near Island Group, is home to a cape that is the westernmost point in the United States. Name this cape.
Cape Wrangell

93. The Brewster Flats are located on what bay in Massachusetts near the Race Point Lighthouse?
Cape Cod Bay

94. The Courthouse and Jail Rocks are in the North Platte River Valley in what Midwestern state?
Nebraska

95. A dam separates Paugus Bay and Opechee Bay in the Lakes Region of what northeastern state bordering Canada?
New Hampshire

96. Pioneer Courthouse Square is located in the downtown area of Portland in what state?
Oregon

97. Augusta, Georgia, is situated on what river formed by the confluence of the Seneca and Tugaloo Rivers?
Savannah River

98. The town of Simmesport is located at the confluence of the Atchafalaya and Red Rivers in what state?
Louisiana

99. The Rio Grande carves out the Santa Elena Canyon in what Texan national park?
Big Bend National Park

100. The Pymatuning Reservoir is located on the border between Pennsylvania and what state to its west?
Ohio

101. Marquette Park is located on Mackinac Island in what state?
Michigan

102. Sweetbay Bogs Preserve is located in what state whose Ross Barnett Reservoir is situated on the Pearl River?
Mississippi

103. Going-to-the-Sun Road is a popular destination in the Rocky Mountains, in what national park in Montana?
Glacier National Park

104. Avalon is a city on Santa Catalina Island in what Californian archipelago?
Channel Islands

105. The Davenport Skybridge is located in what state home to the Rock Island Centennial Bridge?
Iowa

106. The Tug Hill Plateau is an upland region in New York, east of what lake whose name means "Lake of Shining Waters" in the Wyandot language?
Lake Ontario

107. The Tularosa Basin is located east of the Rio Grande in what southwestern state?
New Mexico

108. The Conowingo Dam is located on the Susquehanna River in what state?
Maryland

109. Quinault Rainforest is part of what Washington national park on a peninsula?
Olympic National Park

110. Matagorda Island borders Espiritu Santo Bay on its northern and western sides in what state?
Texas

111. The Natural Chimneys are rock formations in the Shenandoah Valley of what state?
Virginia

112. Mount Marcus Baker is the highest point in what mountain range near the Glenn Highway?
Chugach Mountains

113. The Hells Canyon Wilderness consist of Hells Canyon and the Seven Devils Mountains in what state?
Idaho

114. What major river forms most of the border between South Carolina and Georgia?
Savannah River

115. Mount Elbert is the highest peak in the Rocky Mountains and Colorado. This mountain is in what national forest?
San Isabel National Forest

116. The Belle Isle Conservatory consists of a botanical garden and a greenhouse on the Detroit River, which empties out into what lake?
Lake Erie

117. Name the largest lake entirely within Vermont, in Rutland County.
Lake Bomoseen

118. The Gold Star Memorial Bridge crosses the Thames River between New London and Groton in what state?
Connecticut

119. The Marin Headlands are located north of San Francisco near what famous bridge?
Golden Gate Bridge

120. NorthPark Center is a large shopping mall in what city home to the Reunion Tower?
Dallas

North America

NOTE: This chapter contains questions that can be used to prepare for the **Canadian Geographic Challenge.** However, it can also be useful in the National Geographic Bee for any Canada questions.

1. What country containing the United States Mountain Range is home to Galiano Island, the Tofino district on the Elowista Peninsula, and the city of Espanola on the Spanish River?
   Canada

2. Casa Loma is a museum and landmark located near the CN Tower in Toronto in what country?
   Canada

3. The Nariva Swamp is the largest freshwater wetland in what country where Pitch Lake, the largest natural deposit of asphalt in the world, can be found?
   Trinidad and Tobago

4. The Animal Flower Cave and Harrison's Cave are located in what country whose capital is Bridgetown?
   Barbados

5. San Lorenzo Marine Archipelago National Park is part of the municipality of Mexicali, in the Baja California state of what country?

Mexico

6. Name the lake in southeastern El Salvador that has been declared a Ramsar Wetland.
Lake Olomega

7. Isla de la Juventud is the largest island in the Canarreos Islands. This archipelago, along with the Jardines de la Reina island group, is located in what country?
Cuba

8. Xunantunich, an ancient Mayan archaeological site, is located on a ridge above the Mopan River, in the Cayo District of what country bordering Guatemala?
Belize

9. The Ojo de Liebre Lagoon is in the Mulege Municipality within El Vizcaino Biosphere Reserve in what Mexican state?
Baja California Sur

10. The Monashee Mountains, which border the Arrow Lakes, are near the Selkirk Mountains in what Canadian province?
British Columbia

11. The Carbet Mountains are located in what Caribbean French territory whose highest point is an active stratovolcano known as Mount Pelée?
Martinique

12. Mazinaw Rock, in the Addington Highlands, is located in Bon Echo Provincial Park in what Canadian province?
Ontario

13. The Riberino Zapandi Wetlands are part of a nature reserve in what Costa Rican province?
Guanacaste Province

14. The Cordillera de Talamanca is situated on the border between Costa Rica and what other country?
Panama

15. The Ensenada de la Broa is a bay in southern Cuba and is part of what larger body of water that separates Cuba from Isla de la Juventud?
Gulf of Batabano

16. The fortress of Fortaleza San Felipe in the Puerto Plata Province and the Columbus Lighthouse monument are located in what country?
Dominican Republic

17. The Garibaldi Volcanic Belt is home to Black Tusk Volcano, Garibaldi Lake, and Cinder Cone in what country?
Canada

18. Jiquilisco Bay can be found in the Usulutan Department off the coast of what country?

El Salvador

19. Climbers can explore the ice on the Athabasca Glacier in Alberta in which country?
Canada

20. The Sabana-Camaguey Archipelago, extending to a bay off the coast of Nuevitas, belongs to what country?
Cuba

21. The Papagayo Peninsula, which borders the Gulf of Papagayo, is located in the Guanacaste Province of what Central American country?
Costa Rica

22. Name the Canadian archipelago situated at the confluence of the St. Lawrence and Ottawa Rivers also known as the Montreal Islands.
Hochelaga Archipelago

23. The San Blas Islands and the Pearl Islands are located off the coast of what country bordering the Gulf of Chiriqui to the south?
Panama

24. Lake Peligre, which was constructed because of the Peligre Dam on the Artibonite River, is the second largest lake in what country home to Pic Macaya National Park, in the Massif de la Hotte Mountain Range?
Haiti

25. The Dechibeta Hot Springs can be found in the Limon Province of what country?
Costa Rica

26. Conchaguita is a volcanic island in the Gulf of Fonseca, east of what country?
El Salvador

27. St. John's is a city on what peninsula in eastern Canada?
Avalon Peninsula

28. Guayabo National Monument is situated on the southern slopes of Turrialba Volcano, in the Cartago Province of what country?
Costa Rica

29. Nahanni National Park Reserve is located in the Dehcho Region of what Canadian territory?
Northwest Territories

30. Carbet Falls, located on the Carbet River, is on the lower slopes of La Grande Soufriere in what French territory?
Guadeloupe

31. The Andares Shopping Mall is located in the city of Zapopan in what country?
Mexico

32. Kelowna is a city on what lake in British Columbia that is in a valley of the same name?
Okanagan Valley

33. The Dragon Cliffs are a scenic feature of the Strand Fiord Formation on Axel Heiberg Island in what Canadian territory?
Nunavut

34. Maligne Canyon is located in Jasper National Park in what Canadian province?
Alberta

35. Aridoamerica is a cultural region consisting of parts of what two countries?
United States and Mexico

36. The Sontecomapan Lagoon is located on a plain in what mountain range in Mexico that was used by the Olmec as a source of basalt?
Sierra de los Tuxtlas

37. Uruapan is the second largest city in Michoacan in what country?
Mexico

38. Carrefour is in the Ouest Department, in the southern part of what country?
Haiti

39. Laguna Poco Sol is a volcanic lake in the Tilaran Range of what country home to the Osa Peninsula?
Costa Rica

40. Mount Carleton is located in Mount Carleton Provincial Park in what Canadian province whose Francophone population consists mainly of Acadians?
New Brunswick

41. The Mount Edziza Volcanic Complex is in the Tahltan Highland region of what country?
Canada

42. Saltillo is the capital of what state that is the third largest by area in Mexico?
Coahuila

43. Sumidero Canyon, formed by the Grijalva River, is near the Chicoasen Dam just north of Chiapa de Corzo in what Mexican state?
Chiapas

44. Name the gulf bordering Mexico to the north that shares its name with an isthmus, also to its north.
Gulf of Tehuantepec

45. Dundurn Castle is located in a Canadian city on the Niagara Peninsula. Name this city.
Hamilton

46. The Fort of San Diego is in what Mexican city in the state of Guerrero that has the country's largest beach?
Acapulco

47. What major Central American Lake is located on the border between the Jutiapa and Santa Ana Departments in Guatemala and El Salvador?
Lake Guija

48. The Vinales Valley is a karst area in the Guaniguanico Mountains in the Pinar del Rio Province of what country?
Cuba

49. The Aspotogan Peninsula separates St. Margarets Bay from Mahone Bay in what Canadian province?
Nova Scotia

50. Sulphur Springs, known as the "World's Only Drive-In Volcano", is located in what country whose Pigeon Island is situated in the Gros Islet?
St. Lucia

51. The Tantramar Marshes can be found around the Bay of Fundy on what Canadian isthmus?
Isthmus of Chignecto

52. The Rubén Darío National Theater is in what city near Lake Nicaragua, also known as Lake Cocibolca?
Managua

53. The archaeological sites of Dzibilchaltun, Monte Albán, and Templo Mayor are located in what country bordering the Gulf of Mexico?
Mexico

54. The Arenal Hanging Bridges at Mistico Park can be found in what country?
Costa Rica

55. The Tarascan Plateau is home to Lake Patzcuaro and Lake Cuitzeo in what country home to Paricutin Volcano?
Mexico

56. Sand dunes, lagoons, and salt marshes can be found in Kouchibouguac National Park in what Canadian maritime province?
New Brunswick

57. Boiling Lake, the world's second largest hot lake, is in Morne Trois Pitons National Park in what country?
Dominica

58. What city is known as the "Entertainment Capital" and "Ecotourism Capital" of Honduras and is near Pico Bonito National Park?
La Ceiba

59. Name one of the largest resort towns in the Caribbean Sea, situated on the Hicacos Peninsula in the Matanzas Province, which borders the Bay of Cardenas.

Varadero

60. A part of the Blue Mountains are within Blue and John Crow Mountains National Park. This range is home to Blue Mountain Peak and is the longest in what country?
Jamaica

61. Tufton Hall Waterfall is located just outside the town of Victoria in what country known as the "Island of Spice" and home to Grand Etang Lake?
Grenada

62. The Palace of the Captain Generals is located in what city that was once the capital of the Kingdom of Guatemala and is now a UNESCO World Heritage Site?
Antigua Guatemala

63. What peninsula in Canada lies between Lake Huron and Georgian Bay?
Bruce Peninsula

64. Las Victorias National Park and San Jose la Colonia National Park are located near a Guatemalan city famous for its coffee plantations. Name this city.
Coban

65. The Chinameca Volcano, situated to the north of the San Miguel Volcano, is a stratovolcano in what country?
El Salvador

66. Name the French territory that is one of the Renaissance Islands and is home to Ile Fourchue and Ile Fregate.
St. Barthelemy

67. The Cynthia Peninsula juts out into Lake Temagami in what Canadian province?
Ontario

68. The Columbia Wetlands can be found in the Columbia Valley in what Canadian province?
British Columbia

69. You can find caves and passageways at the Xtabi Cove in what Caribbean country?
Jamaica

70. La Bufadora is a marine geyser on the Punta Banda Peninsula in what Mexican state that occupies Guadalupe Island?
Baja California

71. Name the third largest city in Honduras, in the country's Cortes Department near the Sierra del Merendón Range.
Choloma

72. The Valle de los Ingenios, also known as the Valley of the Sugar Mills, is an UNESCO World Heritage Site just outside what Cuban city in the Sancti Spiritus Province?
Trinidad

73. Name the largest lake by area in Costa Rica.
    Lake Arenal

74. The Siwash Rock is a famous rock formation in Stanley Park in what Canadian city?
    Vancouver

75. The triangular Connaigre Peninsula is located on an island separated from the Labrador Peninsula by the Strait of Belle Isle. Name this island.
    Newfoundland

76. Morne Diablotins, the source of the Toulaman River, is located in what country that is home to the city of Portsmouth, on the Indian River?
    Dominica

77. Banco Chinchorro is an atoll and a reef near Belize in what Mexican state?
    Quintana Roo

78. What city is known as the "Motor City" of Canada, on the Lake Ontario shoreline near Toronto?
    Oshawa

79. Hierve el Agua, a series of rock formations near the Mitla archaeological site, is a natural landmark of what country?
    Mexico

80. The Colorados Archipelago is near the Guanahacabibes Peninsula in what country whose Sierra Maestra Mountain Range is primarily found in the Santiago de Cuba Province?
Cuba

81. Lake Ilopango is a crater lake in the San Salvador, La Paz, and Cuscatlan Departments of what country whose Lake Guija is situated on the border with Guatemala?
El Salvador

82. Darien National Park can be found bordering Los Katios National Park in Colombia, and lies between the Serrania del Darien Mountain Range in what country?
Panama

83. The Guayamouc River flows south from the Massif du Nord in what country home to the Artibonite Valley?
Haiti

84. Name the northernmost point of the Yucatan Peninsula, over thirty miles north of Cancun.
Cabo Catoche (Cape Catoche)

85. Mount Royal is part of the Monteregian Hills in what eastern Canadian city?
Montreal

86. What is the capital and commercial center of the Dutch territory of Sint Maarten?

Philipsburg

87. The Coatepeque Caldera is home to a crater lake of the same name as well as six lava domes, and can be found in what country?
El Salvador

88. Prairieland Park is located in what city that is the most populous in Saskatchewan?
Saskatoon

89. Terraced and stepped turquoise pools are popular features of Semuc Champey Natural Monument in the Alta Verapaz Department of what country?
Guatemala

90. Caja de Muertos is an island near what Puerto Rican city that is known as the "Pearl of the South"?
Ponce

91. Washington Slagbaai National Park is located in the northwestern part of what Dutch territory?
Bonaire

92. Quirigua is an ancient Mayan site along the Motagua River in what Guatemalan department?
Izabal Department

93. Zacaton is a sinkhole near what mountain range in the Mexican state of Tamaulipas?

Sierra de Tamaulipas

94. Bridgetown is located on Carlisle Bay and is the capital of what country?
Barbados

95. Mont Jacques-Cartier is in the Chic-Choc Mountains on what peninsula in Quebec?
Gaspe Peninsula

96. The Lacustrino de Tamborcito Wetlands are part of the Arenal Huetar Norte Conservation Area in the northern part of what country?
Costa Rica

97. Lubaantun is a pre-Columbian city now in the Toledo District, the southernmost region of what country?
Belize

98. Ensenada Honda is an inlet on the northeastern coast of what country home to the Ceiba Municipality and Luis Munoz Marin International Airport?
Puerto Rico

99. The Punta Cometa Peninsula is in what Mexican state with the Guilá Naquitz Cave?
Oaxaca

100. The Sibley Peninsula juts out into Lake Superior in what Canadian province?

Ontario

101. The Samalayuca Dune Fields are composed of quartz and are in the Chihuahua Desert. These dunes are located southeast of Ciudad Juarez in what Mexican state?
Chihuahua

102. What overseas territory of the Netherlands is part of the BES islands along with Sint Eustatius and Saba?
Bonaire

103. The Cerros de Escazú whose highest point is Cerro Rabo de Moco, borders the Central Valley of what country?
Costa Rica

104. Janitzio is an island in Michoacan State, in what lake surrounded by wetlands?
Lake Pátzcuaro

105. Name the Canadian city that is the most populous in New Brunswick, is situated near the Petitcodiac River and Sackville Waterfowl Park.
Moncton

106. The Palisadoes tombolo protects Kingston Harbor, which was previously known as Cagway Bay, in what country?
Jamaica

107. La Romana is located near the island of Santa Catalina in what country?
Dominican Republic

108. Name the largest island in Lake Nicaragua, which formed from two volcanoes rising out of the lake.
Ometepe Lake

109. The Zapata Peninsula can be found east of the Gulf of Batabano and north of the Gulf of Cazones in what country?
Cuba

110. The Sierra de la Minas, separated from the Sierra de Chuacus by the Salama River Valley, is located in what country home to the volcanic lake of Lake Amatitlan?
Guatemala

111. The Chilcotin Group is an area of basaltic lava creating a plateau in what Canadian province?
British Columbia

112. The Mingan Archipelago is located near the Gaspe Peninsula in what gulf?
Gulf of St. Lawrence

113. The Byahaut Bat Cave is located in what country whose capital, Kingstown, is situated on St. Vincent Island?
St. Vincent and the Grenadines

114. The Cathedral of León can be found in the city of León in what country?
Nicaragua

115. Los Tres Ojos is a limestone cave in the municipality of Santo Domingo Este in what country?
Dominican Republic

116. Gran Glaciar Norte, the largest ice cap in Mexico, is on what stratovolcano near a city of the same name?
Pico de Orizaba

117. The Bell Peninsula is on Southampton Island, which is part of what archipelago?
Canadian Arctic Archipelago

118. The Corn Islands are located off the eastern coast of what country whose Solentiname Islands are located in a lake of the same name?
Nicaragua

119. The archaeological site of Bonampak is near Mexico's border with Guatemala in what state?
Chiapas

120. The Pyramid of the Sun and the Pyramid of the Moon were located in what ancient Mesoamerican city now in present day Mexico?
Teotihuacan

121. Punta el Chiquirin, the easternmost point on the El Salvadorian mainland, extends out into what gulf?
Gulf of Fonseca

122. There are over nine hundred indigenous plant species in a wetland in the Matanzas Province of Cuba, on the Zapata Peninsula. Name this wetland.
Zapata Swamp

123. You can see the Burica Peninsula from Volcán Barú in what country?
Costa Rica

124. Desembarco del Granma National Park is located near the Gulf of Guacanayabo and the Jardines de la Reina archipelago in what country?
Cuba

125. The Mer Bleue Conservation Area is located in eastern Ontario east of what major city home to Chaudière Falls and the formerly named Langevin Block?
Ottawa

South America

1. The Araguainha Crater is in Brazil, on the border between the state of Goias and what other state home to Chapada dos Guimaraes National Park, the geographical center of South America?
Mato Grosso

2. The Salinas Grandes is a huge salt flat located at the foot of the Sierras de Cordoba Mountain Range in what South American Cone country?
Argentina

3. Morro do Careca is a large dune in what city in the Brazilian state of Rio Grande do Norte situated on the Potenji River and known as the "City of the Sun"?
Natal

4. Pedra de Sao Domingos is a famous rock formation within the Mantiqueira Mountains in what state that is home to Estadio Independencia in Belo Horizonte?
Minas Gerais

5. La Punta is a wealthy district situated on a province in what city that is the chief seaport of Peru?
Callao

6. Orinduik Falls, on the Ireng River, is in the Pakaraima Mountains on Guyana's border with what country?
   Brazil

7. Punto Fijo is a city on the Paraguana Peninsula, which is connected to Falcon State by the Medanos Isthmus in what country?
   Venezuela

8. Ancohuma is in the northern part of the Cordillera Real and is the third highest mountain in what country?
   Bolivia

9. Punta Galera projects into the Pacific Ocean north of Colun Beach and the Valdivian Coastal Reserve in the Los Rios Region of what country?
   Chile

10. La Mano de Punta del Este is a famous sculpture depicting five human fingers rising above the sand in Punta del Este, near the Laguna del Sauce, in the Maldonado Department of what country?
    Uruguay

11. The Mejillones and Arauco Peninsulas are in what country whose Tres Montes Peninsula is on the Taitao Peninsula and is connected to the Isthmus of Ofqui?
    Chile

12. The "Musical Capital of Venezuela" is noted for its amazing sunsets. Name this city, the capital of the state of Lara.
Barquisimeto

13. Salar de Arizaro, a salt flat in the Andes, is located in what Argentinian province home to Los Cardones National Park?
Salta Province

14. Abrolhos Marine National Park, which is located in the Abrolhos Archipelago, is part of the Central Atlantic Forest Ecological Corridor in what Brazilian state?
Bahia

15. The Macanao Peninsula is a geographical formation on the western end of Margarita Island, the largest island in what Venezuelan state whose capital is La Asuncion?
Nueva Esparta

16. What city in Brazil is located between the Parnaiba and Poti Rivers and is the capital of the state of Piaui?
Teresina

17. The Cerro Chascon-Runtu Jarita complex is a group of lava domes within the Pastos Grandes Caldera in what country?
Bolivia

18. The Wilhelmina Mountains are located in what country whose capital is Paramaribo?
Suriname

19. Orinduik Falls, which lies on the Ireng River, is located in the Pakaraima Mountains on the border between Guyana and what country?
Brazil

20. Kaieteur Falls, the world's largest single drop waterfall by volume, is in Kaieteur National Park on what river?
Potaro River

21. Laguna Colorada is a shallow salt lake and wetland with white-colored and reddish waters. This site is home to populations of James' flamingos and is within Eduardo Avaroa Andean Fauna National Reserve in the Sur Lipez Province of what country?
Bolivia

22. Zapaleri Volcano is located at the tripoint of the borders of Argentina, Bolivia, and what other country home to Almirantazgo Fjord and the Reloncavi Estuary?
Chile

23. Name the strait that connects Winaymarka Lake and Lake Pequeno in Lake Titicaca.
Strait of Tiquina

24. The world's largest cashew tree is a popular destination for many in a city home to the coral formations and crystalline waters of Maracajau. Name this city.
    Natal

25. What city has received a distinction as the "Brazilian Silicon Valley" and is home to Parque Portugal?
    Campinas

26. The Messier Channel, surrounded by Katalalixar National Reserve and Bernardo O'Higgins National Park, is located in what South American Cone country?
    Chile

27. Primera Angostura, a sound lying between the Magallanes Province and Tierra del Fuego Province, is located near the town of Punta Delgada in what country where Pali-Aike National Park can be found?
    Chile

28. Byron Sound, which is home to a sheep farm on Saunders Island, lies northwest of what major island in the Falkland Islands whose largest settlement is Port Howard?
    West Falkland

29. The Carajas Mountains are located within Carajas National Forest in what Brazilian state whose capital, Belem, is situated at the mouth of the Amazon River?
    Para

30. El Junco Lagoon is located on San Cristobal Island in what volcanic archipelago belonging to Ecuador?
Galapagos Islands

31. Kasikasima is a mountain located near Vincent Fayks Airport in the Sipaliwini District of what country?
Suriname

32. The Guiana Space Center is located northwest of the city of Kourou in what overseas department of France?
French Guiana

33. Brazilian Island, located at the confluence of the Uruguay and Quarai Rivers, belongs to what country?
Uruguay

34. The Isla de la Plata is located in Machalilla National Park and is situated off the coast of what country?
Ecuador

35. Chapada Diamantina, also known in Portuguese as the Diamond Plateau, is located in the Espinhaco Mountains and is home to part of the caatinga biome in what Brazilian state?
Bahia

36. What Bolivian city, whose name translates from Spanish to English as "The Heights", is home to a famous art museum that was inaugurated in 2002?

El Alto

37. Name the Peruvian landform that is known to be the world's highest tropical mountain range, and is situated near Lake Conococha.
Cordillera Blanca

38. The Guarani Aquifer is located in the countries of Brazil, Argentina, Paraguay, and what country whose largest body of freshwater is Rincon del Bonete Lake?
Uruguay

39. The Seno Otway, which separates Riesco Island from the Brunswick Peninsula, is connected to the Seno Skyring by the Fitzroy Channel in what country?
Chile

40. Lahuen Nadi Natural Monument is located in the Central Valley of what South American Cone country?
Chile

41. Mainstay Lake is located northwest of the mouth of the Essequibo River in what country where one can find Shell Beach, a popular nesting spot for turtles?
Guyana

42. The Petrohue Waterfalls can be found within Vicente Perez Rosales National Park in what Chilean region whose capital is Puerto Montt?
Los Lagos Region

43. The Guiana Amazonian Park, which was joined with Tumucumaque National Park in Brazil, is the largest French national park and is in what French territory?
French Guiana

44. Ypoa National Park is a protected wetland near the city of Asuncion in what country?
Paraguay

45. Bertha's Beach Important Bird Area is located near Mount Pleasant Airport in what city in the Falkland Islands near Yorke Bay?
Stanley

46. Sol de Manana is a geothermal field in the Potosi Department of what country containing part of the Andean Volcanic Belt?
Bolivia

47. Cayambe is a volcano in the Ecuadorian Andes in what province whose capital and largest city is Quito?
Pichincha Province

48. Runkuraqay is an archaeological site southeast of Qunchamarka in the Cusco Region of what country?
Peru

49. Kukenan tepui is home to Cuquenan Falls near Mount Roraima in Canaima National Park in what Venezuelan region?
Guayana Region

50. The Cordillera de Nahuelbuta is a mountain range along the Pacific Coast of South America and is located in the Bio Bio Region and Araucania Region of what country?
Chile

51. Lake Pilchicocha is located in the Sucumbios Province in what country?
Ecuador

52. Chaupi Orco, the highest peak in the Cordillera Apolobamba, is located in the Puno Region along with Suches Lake in Bolivia and what other country?
Peru

53. Brownsberg Nature Park is located near the Brokopondo Reservoir in what country?
Suriname

54. The Martial Mountains, which are located along the coast of the Beagle Channel, are situated on what island whose eastern portion is in Argentina and whose name means "Fire Land" in Spanish?
Tierra del Fuego

55. What Spanish-speaking city is the highest official capital city in the world and contains the Carondelet Palace?
Quito

56. Punta Ballena is on the southern coast of what country home to the Cuchilla Grande mountain range?
Uruguay

57. Trindade and Martin Vaz is an archipelago located more than seven hundred miles east of the city of Vitoria, the capital of what Brazilian state?
Espirito Santo

58. The seat of the Constitutional Court of Peru is in a city known for its Basilica Cathedral and its close proximity to Misti Volcano. Name this city.
Arequipa

59. The Seven Lakes begin at Lacar Lake in the Neuquen Province of Argentina and end at Corral Bay at the mouth of the Valdivia River in what country?
Chile

60. Taulliraju Mountain can be found within Huascaran National Park in what country?
Peru

61. The Kanuku Mountains, which are located in the Upper Takutu-Upper Essequibo Region, can be found in what country with the Rupununi River?

Guyana

62. Antisana is a stratovolcano in the Andes Mountains and is located near what city in Ecuador?
Quito

63. The Caverna de Tapagem is located within Jacupiranga State Park in what Brazilian state that is the wealthiest in the country and is known as the "Locomotive of Brazil"?
Sao Paulo

64. The San Francisco Glacier is in El Morado Natural Monument and is part of the Volcan River Basin in what country?
Chile

65. Mount Maria is located in the Hornby Mountains of what archipelagic territory?
Falkland Islands

66. Qalsata is a mountain in the Laraceja Province and is situated east of San Francisco Lake in what country home to the Arbol de Piedra rock formation?
Bolivia

67. Salar de Surire Natural Monument lies south of the stratovolcano of Arintica in what country?
Chile

68. Saint-Laurent-du-Maroni, which is situated on the Maroni River, is a commune of what dependent territory?
French Guiana

69. The Araya Peninsula, which juts out into the Caribbean Sea, is located in the Sucre State of what country?
Venezuela

70. Yacyreta Island, on the Parana River, is near Yacyreta Dam, a hydroelectric power plant between the Corrientes Province in Argentina and what Paraguayan city in the Misiones Department?
Ayolas

71. Huishue Lake is near Maihue Lake and Puyehue-Cordon Caulle, a mountain massif in Puyehue National Park in what country?
Chile

72. The Pantanal, home to the world's largest tropical wetland area and Pantanal Matogrossense National Park, borders the Chiquitano Dry Forests. This national park is in Mato Grosso and what other Brazilian state?
Mato Grosso do Sul

73. La Portada Natural Monument, a natural arch off the coast of Chile, is located near what major city home to Andres Sabella Galvez International Airport and the Tropic of Capricorn Monument?
Antofagasta

74. The Yampupata Peninsula, located in the Manco Capac Province of the La Paz Department in Bolivia, juts out into what lake?
Lake Titicaca

75. The Altiplano Cundiboyacense is a plateau in the Cordillera Oriental in what country home to Tequendama Falls?
Colombia

76. The Merida Cable Car, the highest cable car in the world, is located in what Venezuelan state where Pico Bolivar can be found?
Merida

77. Llovizna Falls is a scenic area situated near the Macagua Dam in what Venezuelan city?
Ciudad Guayana

78. Saltos del Monday is a famous waterfall in the Alto Parana Department of Paraguay near what Paraguayan city in the Triple Frontier?
Ciudad del Este

79. Cotopaxi is a famous stratovolcano in the Cotopaxi Province of what country with the city of Latacunga?
Ecuador

80. The city of La Paz can be found in a canyon carved out by what river whose source is near Mount Chacaltaya?
Choqueyapu River

81. Presidente Rios Lake lies in the center of what Chilean peninsula in the Aysen Region?
Taitao Peninsula

82. Duque de Caixas is on what bay in the state of Rio de Janeiro that is the second largest in Brazil?
Guanabara Bay

83. The Huaytapallana Mountains are located in what country home to Mejia Lagoons National Sanctuary and Lake Salinas?
Peru

84. Keppel Sound, which contains Golding Island and Keppel Island, is located in what English territory home to the Lafonia Peninsula?
Falkland Islands

85. Pongo de Mainique is in the Vilcabamba Mountain Range and divides the Urubamba River in what country?
Peru

86. What city is in the highly developed region of Triangulo Mineiro and is the second largest city in Minas Gerais?
Uberlandia

87. El Barco Lake is located near Callaqui, a volcano found in Ralco National Reserve in what country?
Chile

88. Jaboatao dos Guararapes is an important industrial center and is located within the Recife Metropolitan Area in what country?
Brazil

89. The Serrania de Macuira is located in the La Guajira Desert on the La Guajira Peninsula in what country home to Morrocoy National Park?
Venezuela

90. Name the world's largest salt flat, a major breeding ground for flamingos and located in the Potosi Department of Bolivia.
Salar de Uyuni (Salar de Tunupa is acceptable)

91. The Sierra de Famatina is a mountain range in what province in Argentina home to Talampaya National Park?
La Rioja Province

92. The Paracas Peninsula is located within Paracas National Reservation in what country?
Peru

93. The Guia Narrows, which connect the Sarmiento Channel with the Inocentes Channel, border Hanover Island and Chatham Island in what country?

Chile

94. The Chongon Colonche Range is located in the provinces of Guayas and Manabi in what country?
Ecuador

95. The Monturaqui Crater is located south of the Salar de Atacama, a Chilean salt flat in what region comprising three provinces?
Antofagasta Region

96. What Argentinian city is connected to Buenos Aires by bridges crossing over the Matanza River?
Avellaneda

97. Cajas National Park can be found near the city of Cuenca, the capital of what Ecuadorian province?
Azuay Province

98. The Moraleda Channel separates the Chonos Archipelago from what country bordering the Gulf of Corcovado?
Chile

99. The Juan Carlos Castagnino Municipal Museum of Art is in what Argentinian city served by Astor Piazzolla International Airport?
Mar del Plata

100. Part of the Vizcachas Mountains lie in La Campana National Park in what Chilean region that administers Easter Island?
Valparaiso Region

101. The Darien Gap is a break in the Pan-American Highway and is located in Panama and the Choco Department of what country?
Colombia

102. The source of the Putumayo River is east of Pasto, a Colombian city at the foot of what volcano?
Galeras Volcano

103. Lake Valencia is a body of water outside what city in Venezuela where Arturo Michelena International Airport can be found?
Valencia

104. Cano Cristales, known as the "River of Five Colors", is referred to as the most beautiful river in the world and is in what country?
Colombia

105. The Imataka Mountains are home to large deposits of iron ore and are located in the northwestern part of what country?
Guyana

106. Rapa Nui National Park is located on what Chilean-controlled island?
Easter Island

107. Choquequirao is an Incan archaeological site in the Vilcabamba Mountains of what country?
Peru

108. The Cerro de Arcos is a rock formation on the border between the El Oro and Loja Provinces in what country?
Ecuador

109. The Smyth Channel, whose northern entrance is in the Nelson Strait, is located in what country home to the Munoz Gamero Peninsula?
Chile

110. Name the Peruvian city on the banks of the Moche River from where the country's judiciary originated.
Trujillo

111. Cape Pembroke, the easternmost point in the Falkland Islands, is on what island bordering Choiseul Sound?
East Falkland

112. The Magdalena Channel borders Aracena Island and passes through Alberto de Agostini National Park in what country?
Chile

113. Name the ecoregion of Brazil made up of tropical savanna that includes gallery forests and extends partially into Paraguay and Bolivia.
Cerrado

114. Caral is an archaeological site that was once part of the Norte Chico Civilization and is now in the Barranca Province of what country?
Peru

115. Comodoro Rivadavia, situated on San Jorge Gulf at the base of Chenque Hill, is a city in what country?
Argentina

116. You can visit many parks and the Estadio Monumental de Maturin in what Venezuelan city?
Maturin

117. The capital of the Paraiba State in Brazil is also the easternmost city in the Americas. Name this city.
Joao Pessoa

118. Santander Park is located in the city of Cucuta, in what country?
Venezuela

119. Lagoon Mirim is connected to the Lagoa dos Patos, the largest barrier lagoon in South America, by what channel?
Sao Goncalo Channel

120. The Las Lajas Sanctuary is a popular destination in the Guaitara River Canyon, in the city of Ipiales in what country?
Colombia

121. Banos is a city surrounded by hot springs and is situated near Tungurahua Volcano. This city, known as the "Gateway to the Amazon", is located in what country?
Ecuador

122. Raquira, a city known for its exquisite pottery, can be found near Lake Fuquene in what country?
Colombia

123. The geyser field of El Tatio can be found in what country known for its Camanchaca cloud banks and Lauca National Park?
Chile

124. The Beni River joins the Mamore River at the border between Brazil and what country?
Bolivia

125. El Ateneo Grand Splendid is renowned to be one of the most beautiful bookshops in the world, and can be found in what Argentinian city home to the Palermo Woods?
Buenos Aires

126. Saksaywaman is a citadel on the outskirts of what famous ancient city located near the Urubamba Valley?
Cusco

127. The volcanic crater of Rano Raraku is situated on the lower slopes of what volcano?
Terevaka Volcano

128. The Juruena River originates in the Parecis Plateau in what Brazilian state?
Mato Grosso

129. The "City of Parks" is situated in the Cordillera Oriental of Colombia and is home to Alfonso Lopez Stadium. Name this city.
Bucaramanga

130. Water-filled Quilotoa is a caldera near the city of Latacunga in what Ecuadorian province?
Cotopaxi Province

131. Tayrona National Natural Park is located near the city of Santa Marta, which was the first Spanish settlement in what present day country?
Colombia

132. The Niterio Contemporary Art Museum is one of the landmarks of Niteroi, a city in what Brazilian state?
Rio de Janeiro

133. The pristine coral reefs are an amazing sight in an archipelago 80 miles north of La Guaira. Name this archipelago, declared a national park by the Venezuelan government in 1972.
Los Roques

134. The Essequibo River originates in the Acarai Mountains near what country's border with Brazil?
Guyana

135. Colorful stone formations cover the landscape of Valle de la Luna, a valley that is part of Los Flamencos National Reserve in what country?
Chile

136. White sand dunes and rainwater lagoons make up Lencois Maranhenses National Park in what Brazilian State?
Maranhao State

137. Name Bogota's most populous suburb, whose name is derived from the Chibcha language.
Soacha

138. The Bermejo and Paraguay Rivers meet north of Resistencia, a major city in what country?
Argentina

139. What Chilean city lies near Osorno Volcano and Puyehue National Park and can be found in the Los Lagos Region?

Osorno

140. The city of Londrina is located in the Parana State of what country?
Brazil

141. Villavicencio is on the banks of the Guatiquia River and is the capital of the Meta Department of what country?
Colombia

142. Barcelona is the capital of the Anzoategui State, a place well known for its beaches in what country?
Venezuela

143. The city of Tacna is situated in the Caplina River Valley and is home to the Toquepala Caves in what country?
Peru

144. Monteria, connected to the Caribbean Sea by the Sinu River, is home to Los Garzones Airport and is a city in what country?
Colombia

145. Machala, referred to as the "Banana Capital of the World", is in the lowlands near what Ecuadorian gulf?
Gulf of Guayaquil

146. The source of the Tiete River is in the Salesopolis Municipality in what country?
Brazil

147. The city of Puerto Montt is situated on Reloncavi Sound near the Argentinian Nahuel Huapi Lake in what country?
Chile

148. Copiapo is the capital of the Copiapo Province and the Atacama Region of what country?
Chile

149. Cajamarca, known for its dairy products, was the site where Spanish invaders defeated what empire?
Inca Empire

150. The "Coffee Axis", also renowned as the Coffee Triangle, is home to a type of coffee considered by many as one of the best in the world. This famous region is part of the Paisa Region of what country?
Colombia

Asia

1. Namaste Falls is located in the Koshi Zone in what country home to the Pokhara Valley?
    Nepal

2. Cape Sukhoy Nos is located on what island in the Novaya Zemlya Archipelago separated from Yuzhny Island to the south by the Matochkin Strait?
    Severny

3. The Tajhat Palace, home to the Rangpur Museum, is located in the city of Rangpur in what country home to Dhanmondi Lake?
    Bangladesh

4. What country's longest river, the Mahaweli River, has its source in Horton Plains National Park?
    Sri Lanka

5. Runners travel about 150 miles in a desert race called the Gobi March in which Asian country?
    China

6. The Khorat Plateau, divided by the Phu Phan Mountains, is drained by the Mun and Chi Rivers in the Isan Region of what country?

Thailand

7. Carstensz Glacier is near Grasberg Mine, the world's largest gold mine and third largest copper mine. These two features are located in what archipelagic country?
Indonesia

8. The Ha Tien Islands and the Ba Lua Islands are in the Gulf of Thailand and belong to what country home to the Bay Nui Mountains, close to its border with Cambodia?
Vietnam

9. The Saimaluu Tash is a petroglyph site home to a huge collection of rock art. This site is located in the Jalal-Abad Province, near the village of Kazarman in what country home to the Boom Gorge and Song Kol Lake?
Kyrgyzstan

10. Seongsan Ilchulbong, also known as "Sunrise Peak", is a unique landform located in what country home to Seoraksan, the highest peak in the Taebaek Mountains?
South Korea

11. Lake Kamyslybas is located in the Kyzylorda Province of what country whose Lake Sasykkol is part of Alakol Biosphere Reserve?
Kazakhstan

12. The Alagalla Mountains are at the border between the Central Province and the Sabaragamuwa Province of what country?
Sri Lanka

13. Sharyn Canyon, carved out by the Sharyn River, is east of Almaty in what Kazakh region?
Almaty Region

14. Lake Terkhiin Tsagaan, in the Khangai Mountains, is part of what country whose largest freshwater lake is Khovsgol Nuur?
Mongolia

15. The Kharan Desert is located in the Balochistan Province of what country?
Pakistan

16. The Azraq Wetland Reserve, an oasis for migratory birds, is located near the city of Azraq in what country?
Jordan

17. The Kopet Dag is a mountain range on the border between Iran and what other country home to the Krasnovodsk Peninsula and the Cheleken Peninsula?
Turkmenistan

18. Rara Lake, located in Rara Lake National Park, is the largest and deepest freshwater lake in what landlocked country home to the Kali Gandaki Gorge?

Nepal

19. The Zamboanga Peninsula, situated between Moro Gulf and the Sulu Sea, is on what Philippine island?
Mindanao

20. The Illi River and Dzungarian Altai Mountains can be found in what country?
Kazakhstan

21. The peak of Manucoco is located on Atauro Island, which is situated at the western end of the Wetar Strait in what country?
East Timor (Timor-Leste)

22. Fujiyoshida is famous for its udon noodles, and can be found at the base of what stratovolcano in Japan?
Mount Fuji

23. The Bumdeling Wildlife Sanctuary, home to the former Kulong Chu Wildlife Sanctuary, is in what country?
Bhutan

24. Mount Korbu, situated in the southern Tenasserim Hills, is the highest peak in what mountain range in Malaysia situated to the west of Kenyir Lake?
Titiwangsa Mountains

25. Tianchi is an alpine lake in the Bogda Shan, a mountain range in what autonomous region of China?

Xinjiang Uyghur Autonomous Region (Xinjiang Uyghur or Xinjiang)

26. The Ara Canal, connecting the Han River to the Yellow Sea, is in South Korea and extends from Seoul to what city with Bupyeong Station?
Incheon

27. The Kishon River feeds Haifa Bay in what country?
Israel

28. What Southeast Asian country's Pak Ou caves overlook the Mekong River?
Laos

29. The Al-Ghab Plain, which separates the Syrian Coastal Mountain Range and the peak of Jabal Zawiya, is located in what country home to the castle of Krak des Chevaliers, a World Heritage Site?
Syria

30. The Fedchenko Glacier, the longest glacier in the world outside a polar region, is located in the Yazgulem Range of the Pamir Mountains, in the Gorno-Badakhshan Province of what country?
Tajikistan

31. The Muyunkum Desert extends to the Karatau Mountains in the Jambyl Region of what country?
Kazakhstan

32. Mount Hamiguitan is a World Heritage Site in the Davao Oriental Province of what country?
Philippines

33. The Hakone Botanical Garden of Wetlands is located in Fuji-Hakone-Izu National Park and is near the Fuji Five Lakes. This garden is in what Japanese prefecture?
Kamagawa Prefecture

34. The Kackar Mountains, whose highest peak is Kackar Dagi, are the highest part of what Turkish mountain range home to the Sumela Monastery?
Pontic Mountains

35. The peak of Kongur Tagh is situated near Muztagh Ata and Karakul Lake, which in turn is located approximately 120 miles from the city of Kashgar in what country?
China

36. Badab-e Surt is a site of stepped travertine terrace formations created over thousands of years by water from mineral hot springs and carbonate minerals. This site is located about sixty miles south of Sari, the capital of the Mazandaran Province of what country?
Iran

37. The Chom Ong Caves, in Oudomxay Province, are near the province's capital of Muang Xay, in what country?
Laos

38. What Sri Lankan national park is known for its migratory seabirds and is contiguous with Yala National Park?
Kumana National Park

39. Lake Borovoe is located within Burabay National Park near the city of Shchuchinsk in what Kazakhstani region?
Akmola Region

40. The La Paz Sand Dunes are located near Laoag, known as the "Sunshine City". This city is the capital of the Ilocos Norte Province in what country?
Philippines

41. Jusangjeolli Cliff is a volcanic rock formation on what South Korean island?
Jeju Island

42. The Israeli city of Haifa is home to the Baha'i World Center on the slopes of what mountain?
Mount Carmel

43. The Dashti Salt Dome is located in the Bushehr Province in what Iranian mountain range?
Zagros Mountains

44. Snake Gorge is in the Ad Dakhiliya Region of Oman, near the city of Nizwa. Nizwa is at the foot of the Hajar Mountains in what country?
Oman

45. Lake Tengiz is a wetland ecosystem important for hundreds of bird species and is part of the Korgalzhyn Nature Reserve in two provinces of what Central Asian country?
Kazakhstan

46. Porak is a stratovolcano located near Lake Sevan on the border between Armenia and what other country whose Lake Sarysu is located along the Kura River?
Azerbaijan

47. The Tarbagatai Mountains in eastern Kazakhstan border what lake whose only outflow is the Irtysh River and is the largest lake in the East Kazakhstan Province?
Lake Zaysan

48. The Trashiyangtse District is part of what country home to the Phobjikha Valley and Gangteng Monastery?
Bhutan

49. The Duung Wetland, a coastal sand dune, is located in the South Chungcheong Province of what country?
South Korea

50. Bishazari Tal is an oxbow lake system between the Siwalik Hills and the Mahabharat Range in what country?
Nepal

51. Botum Sakor National Park is located near Sihanoukville and is home to mangrove forests in what country?
Cambodia

52. The Caspian Hyrcanian Mixed Forests ecoregion can be found in the Lerik District in Azerbaijan and areas in the northern part of what large country?
Iran

53. Haba Snow Mountains is located near Tiger Leaping Gorge in what mountain range that includes the Chola Mountains?
Shaluli Mountains

54. What Sri Lankan bay is located on the southeastern coast of Sri Lanka and is situated south of the market town of Pottuvil?
Arugam Bay

55. Huishan National Forest Park, which is adjacent to Xihui Park, is located near the Grand Canal. This protected area is located in what Chinese province?
Jiangsu Province

56. The Qandil Mountains are located near what country's border with Iran?
Iraq

57. The Kebar Valley is situated on the Bird's Head Peninsula and borders the Tamrau Mountains in the province of West Papua in what country?
Indonesia

58. The Taurus Mountains extend from Lake Egirdir, in what Turkish province home to the Yivliminare Mosque and the Duden Waterfalls?
Antalya Province

59. The Pearl-Qatar is an artificial island situated near the West Bay Lagoon area of what city in Qatar home to the marketplace of Souq Waqif?
Doha

60. Khao Phing Kan is an island group consisting of limestone tower karsts, such as Ko Tapu. These islands are located in Phang Nga Bay in Ao Phang Nga National Park, in the Phang Nga Province of what country?
Thailand

61. The Yandang Mountains are a coastal range home to the Lingfeng Peaks in what Chinese province whose capital is Hangzhou?
Zhejiang Province

62. The Trus Madi Range is located in what Malaysian state home to Crocker Range National Park?
Sabah

63. The Oriental Mindoro Province borders the Verde Island Passage in what country?
Philippines

64. The Siachen Glacier, which is controlled by India, is the longest glacier in the Karakoram Range and is claimed by what other country?
Pakistan

65. Shanidar Cave is an archaeological site on the mountain of Bradost in what country home to Bekhme Dam, which is situated on the Great Zab River?
Iraq

66. The Gurvan Saikhan Mountains form the eastern part of Gobi Gurvansaikhan National Park. This area is home to the Khongoryn Els, popularly known as the "Singing Sands", which are in what country?
Mongolia

67. Dondra Head is a cape on the southern tip of what country home to Casuarina Beach, a popular attraction near the city of Jaffna?
Sri Lanka

68. The Hazarchishma Natural Bridge, the world's twelfth largest natural bridge, is located in what country home to the Bamiyan Province, in the Hazarajat Region?
Afghanistan

69. Tiga Island is located within Tiga Islands National Park and is situated near Tunku Abdul Rahman National Park off the coast of Sabah in what country?
Malaysia

70. The Khaju Bridge is located across the Zayandeh River in what Iranian province home to Naqsh-e Jahan Square, one of the largest squares in the world?
Isfahan Province

71. The Sinjar Mountains, which are considered sacred by the Yazidi people, can be found in what country?
Iraq

72. Bosten Lake is situated on the northeastern edge of what basin whose historical name was Altishahr and is located near the Yumen Pass and the Gansu Corridor?
Tarim Basin

73. Tubbataha Reefs Natural Park is located southeast of Puerto Princesa, a city in what province in the Philippines in the Mimaropa Region?
Palawan Province

74. The Siypantosh Rock Paintings are located throughout the southwestern part of the Zarafshan Mountains, in the Qashqadaryo Region of what country home to the Aralkum Desert and part of the Fann Mountains?
Uzbekistan

75. The Dieng Plateau is a marshy area straddling the Dieng Volcanic Complex in what country home to Puru Besakih, a Hindu temple complex on the slopes of Mount Agung?
Indonesia

76. Name the administrative division of Kazakhstan whose capital is Kokshetau and contains Lake Kopa.
Akmola Region

77. In the Philippines, the Dinagat Islands are separated from the island of Leyte by what strait?
Surigao

78. The Hanhowuz Reservoir is part of the Karakum Canal system in the Mary Region of what country?
Turkmenistan

79. The Muthurajawela Wetlands and Kirala Kelle Wetlands are located in what country home to Bundala National Park, harboring populations of greater flamingos?
Sri Lanka

80. What country is home to the Mai Pokhari Wetland, a pilgrimage site for both Hindus and Buddhists?
Nepal

81. Cape Inamuragasaki is located in what Japanese city home to the artificial Seven Entrances and situated on the Nameri River in the Kanto Region?
Kamakura

82. The Yarkon River passes through what city that is the most populous in Israel's Gush Dan region?
Tel Aviv

83. Jinmu Cape, considered the southernmost point in China, is part of Sanya City, on what island in China that is separated from the Leizhou Peninsula by the Qiongzhou Strait?
Hainan Island

84. Name the peninsula that is the easternmost point of the New Territories region of Hong Kong.
Sai Kung Peninsula

85. The Wahiba Sands, now known as the Sharqiya Sands, are located in what country home to the Al Hajar Mountains, situated next to the Nakhal Fort?
Oman

86. Gotjawal Forest is on Hallasan, the tallest mountain in South Korea, in Hallasan National Park on what island?
Jeju Island

87. The region of Cukurova, known by its historical name of Cilicia, borders the Gulf of Iskenderun in what country?
Turkey

88. The Zinghmuh Mountains are located in the Chin Hills, in the Chin State of what country whose northernmost national park is Doi Pha Hom Pok National Park?
Myanmar

89. The Burgan Oil Field is located in what country home to the Arraya Tower and the Mutla Ridge?
Kuwait

90. Name the largest island in Turkmenistan, located in the Caspian Sea, belonging to the country's Balkan Region.
Ogurja Ada

91. The Nineveh Plains are home to many ancient religious sites, including the Mar Mattai Monastery. This monastery is atop Mount Alfaf in what country?
Iraq

92. The Jezreel Valley borders the Mount Carmel Range to the west and the Jordan Valley to the east. This plain in the Lower Galilee Region is located in what country?
Israel

93. In what province in China can you visit the Beisi Pagoda and the Chaotian Palace?
Jiangsu Province

94. The Mugodzhar Hills are located in the Aktobe Region of what country?
Kazakhstan

95. The Bimmah Sinkhole is located in the eastern part of the Muscat Governorate in what country?
Oman

96. The Jeita Grotto is a system of karst limestone caves situated in the Nahr al-Kalb River Valley. This cave system, inhabited during prehistoric times, is a national symbol of what country?
Lebanon

97. The Greater and Lesser Tunbs are two islands in the Persian Gulf claimed by the United Arab Emirates but administered by what country's Hormozgan Province?
Iran

98. The Shimosa Plateau, on the Kanto Plain, is home to Narita International Airport in what Japanese city where you can see Mount Kumotori, in the Okuchichibu Mountains?
Tokyo

99. The Cu Lao Re Volcanic Islands are located northeast of the city of Quang Ngai in what country whose Khau Pha Pass is home to rice terraces?
Vietnam

100. The Ryn Desert is southeast of the Volga Upland in what country?
Kazakhstan

101. The Belen Pass, which is located in the Nur Mountains, is also known as the Syrian Gates and is situated in the Hatay Province of what country home to Kapikaya Canyon and Saklikent National Park?
Turkey

102. You can ride a ferry from the Yuantouzhu Peninsula to the Sanshan Islands in what lake?
Lake Tai

103. The Golden Age Lake is currently under construction in the Karashor Depression of what country?
Turkmenistan

104. The Judaean Mountans' northern region is known as the Samarian Hills. Samaria is a mountainous region of the ancient Eastern Mediterranean, in the northern part of what country?
Israel

105. What Eastern Mediterranean country is home to Palm Islands Natural Reserve?
Lebanon

106. Rasht is the capital of the Gilan Province and was historically known as the "Gate of Europe". This city is located on what country's Caspian Sea coast?
Iran

107. The Manora Peninsula is connected to the mainland by the Sandspit in what country?
Pakistan

108. The Jeti-Oguz Rocks are unique rock formations located in the Issyk-Kul Province of what country that shares Lenin Peak with Tajikistan?
Kyrgyzstan

109. The Javakheti Range, whose highest point is Mount Achkasar, is located in what country home to Lake Arpi, the source of the Akhurian River?
Armenia

110. Khong Island is the largest island in the Si Phan Don Archipelago, which is situated in the Mekong River in the Champasak Province of what country?
Laos

111. The Straits of Tiran, which separate the Sinai Peninsula from the Arabian Peninsula, border what country to the east?
Saudi Arabia

112. Ratargul Swamp Forest is located in what country home to Kaptai Lake, which was formed by creating the Kaptai Dam on the Karnaphuli River?
Bangladesh

113. The Ngong Ping Highland is home to the Po Lin
     Monastery and the Tian Tan Buddha statue on Lantau
     Island in what country?
     China

114. The Kaydak Inlet, which forms the eastern limit of the
     Buzachi Peninsula, borders Dead Kutluk Bay, near
     Durneva Island in what country?
     Kazakhstan

115. Thousands of stone jars make up the Plain of Jars, on
     what Laotian plateau?
     Xiangkhoang Plateau

116. The Tabun-Khara-Obo Crater is located in the Dornogovi
     Province of what country home to the Khentii Mountains
     and Burkhan Khaldun, an UNESCO World Heritage Site?
     Mongolia

117. Sigiriya is an ancient rock fortress and World Heritage
     Site located near Dambulla in the Central Province of
     what country?
     Sri Lanka

118. The capital of Iran's East Azerbaijan Province is home to
     the largest covered bazaar in the world, along with the
     park of El-Golu. Name this city, known as the "World's
     Carpet and Crafts City".
     Tabriz

119. The Iori Plateau, which is located between the Kura River and Alazani River, is occupied in its eastern part by the Shiraki Plain in what country?
Georgia

120. The oasis city of Merv is on the Murghab River and is home to the Tomb of Ahmed Sanjar in what country?
Turkmenistan

121. The Yangpu Peninsula can be found in the northwestern part of what Chinese island?
Hainan

122. Jabal an Nabi Shu'yab is the highest point on the Arabian Peninsula. This mountain is located in the Sanaa Governorate of what country?
Yemen

123. The Geli Ali Beg Waterfall is located in the Kurdistan Region of what country home to the Bekhal Waterfall and the Shahrizor Plain?
Iraq

124. The Donggung Palace and Wolji Pond are located in Gyeongju National Park in North Gyeongsang Province in what country?
South Korea

125. Naujan Lake, near the city of Calapan, is on the island of Mindoro in what Filipino province?

Oriental Mindoro Province

126. Name the longest inland river in China, having its origins near Aral in Xinjiang.
Tarim River

127. The Atsumi Peninsula, whose climate is affected by the Kuroshio Current, faces the Chita Peninsula. This peninsula, whose coastal areas are part of Mikawa-wan Quasi-National Park, is located in what country?
Japan

128. The Dapeng Peninsula, which borders Daya Bay and Mirs Bay, is home to the Dapeng Fortress in what country?
China

129. The Zhinvali Dam on the Aragvi River generates much of what country's power?
Georgia

130. The Sarugamori Sand Dunes are located on the Shimokita Peninsula, which is home to the Yagen Valley. This valley is in Shimokita Hanto Quasi-National Park in what country?
Japan

131. The Acarlar Floodplain Forest is located on the coast of the Black Sea in the Sakarya Province of what Mediterranean country?
Turkey

132. The Kaema Plateau, which is surrounded by the Rangrim Mountains, is located in what country home to the Songam Cavern and Yanggakdo, an island on the Taedong River?
North Korea

133. Mount Putuo, whose name is derived from the Sanskrit name "Potalaka", is considered one of the four sacred mountains in Chinese Buddhism. This mountain is situated in the Zhoushan Archipelago, southeast of Shanghai in what Chinese province?
Zhejiang Province

134. Lake Yeak Laom is located near Banlung, the capital of Ratanakiri Province of what country?
Cambodia

135. What country's Selat Pandan Strait separates its Jurong Island from its Southern Island?
Singapore

136. The Timan Ridge lies west of the Pechora River, and Yugyd Va National Park lies to the east. This protected area, in the Komi Republic, is the second largest national park in what country?
Russia

137. The Bintang Mountains are part of what larger mountain range in Malaysia spanning three countries?

Tenasserim Hills

138. Korgalzhyn Nature Reserve is part of an UNESCO World Heritage Site in what upland region of Kazakhstan?
Kazakh Uplands

139. The Khatanga River creates an estuary as it flows into the Khatanga Gulf of what sea?
Laptev Sea

140. Umm Qasr is a port city on the Khawr Abd Allah Estuary and is located on the Al-Faw Peninsula in what country bordering the Persian Gulf?
Iraq

141. Vaadhoo Island is known for its bioluminescent "sea of stars", which shines bright blue due to sea plankton. This island is located in what archipelagic country?
Maldives

142. What country is home to King Fahd International Airport and King Fahd's Fountain?
Saudi Arabia

143. The Dalton Pass is a zigzagged road that is also an entrance to the Cagayan Valley, an administrative region consisting of five provinces of what country?
Philippines

144. Poti is a port city near the ancient Greek settlement of Phasis and is surrounded by Kolkheti National Park in what country?
Georgia

145. Nong Fa Lake, a volcanic crater lake located within the Dong Ampham National Biodiversity Conservation Area, is located in the Attapeu Province of what country?
Laos

146. The Foca Islands, whose largest island is Uzanada, are in the Gulf of Izmir and belong to what country whose Akyatan Lagoon is located in the Cukurova Region?
Turkey

147. Tabriz is located north of Sahand Volcano and is home to the Mausoleum of Poets in what country?
Iran

148. Jurong Lake is a freshwater lake and reservoir in Jurong East, a town and planning area in what country?
Singapore

149. Shuklaphanta Wildlife Reserve is contiguous with India's Kishanpur Wildlife Sanctuary and is located in the Far-Western Development Region of what country?
Nepal

150. Name the highest peak in the Lebanon Mountains and Lebanon.

Qurnat as' Sawda

151. The Pasupati Bridge and the Merdeka Building are located in what Indonesian city?
Bandung

152. The Tyuleniy Archipelago is located off the northern coast of the Mangyshlak Peninsula and can be found in the Mangystau Region of what country?
Kazakhstan

153. Todoroki Athletics Stadium is used mostly for football matches and is located in what city home to the Lazona Kawasaki Plaza and situated on the Tama River?
Kawasaki

154. The San Roque Cathedral and the Bonifacio Monument are historical landmarks in what Philippine city?
Caloocan

155. The Khor Kalmat Lagoon along the Makran coast of what country?
Pakistan

156. The Myeongryang Strait separates Jindo Island from the mainland of what country?
South Korea

157. The Temple of Hercules and Umayyad Palace are famous landmarks in the Amman Citadel in what country?

Jordan

158. What country's Lachin Corridor forms the shortest route between Armenia and Nagorno-Karabakh?
Azerbaijan

159. The Sakhir Desert is an arid area near the village of Al Zallaq in what country?
Bahrain

160. The Klang Valley borders Rawang, once a major tin-producing center, in what country?
Malaysia

161. The largest island in the Wallacea geographical region is what island west of the Maluku Islands?
Sulawesi

162. A city on the Pothohar Plateau, near Benazir Bhutto International Airport and Ayub National Park, became a temporary capital of Pakistan in 1959. Name this city.
Rawalpindi

163. The Yancheng Coastal Wetlands, which are located near the Songnen Plain, are located in what Chinese province whose largest lake is Lake Tai?
Jiangsu Province

164. The Chindwin River originates in what valley in the Kachin State of Myanmar?

Hukawng Valley

165. The Kacchi Plain is located in the southern part of what Pakistani province home to Astola Island?
Balochistan

166. The Bukit Peninsula, home to the Uluwatu Temple, is located on what Indonesian island with a Hindu majority?
Bali

167. Shandur Top, known as the "Roof of the World", is near Rakaposhi Mountain in what country's Spin Ghar Range?
Pakistan

168. The Rosh HaNikra Grottoes are chalk cliff faces that open up into multiple grottoes and is a geologic formation situated near Nahariya, the northernmost coastal city in what country?
Israel

169. The Vyatka River is in the Republic of Tatarstan and is a tributary of what river that has its mouth in the Volga River?
Kama River

170. What country's Well of Zamzam is located in the Masjid al-Haram, the world's largest mosque?
Saudi Arabia

171. The Khazar Islands are a planned group of artificial islands also known as the Caspian Islands. This initiative, scheduled to be completed between 2020 and 2025, will be located in what country?
Azerbaijan

172. The Armutlu Peninsula, which separates the Gulf of Izmit from the Gulf of Gemlik, is located in the Yalova, Kocaeli, and Bursa Provinces of what country?
Turkey

173. The mangrove forests of Qeshm are located on what country's southern coast?
Iran

174. The Flaming Mountains are dry red sandstone hills that have eroded over time, and are in the northern part of the Taklamakan Desert, in the Tian Shan Mountain Range of what country?
China

175. The Ghaggar-Hakra River flows through the Cholistan Desert, whose name is derived from the Turkic word *chol*, meaning desert. This river is in what country?
Pakistan

176. Name the largest freshwater lake in Iran, in the country's Fars Province.
Lake Parishan

177. Name the bay that can be found in Qingdao, straddling a long sea bridge of the same name.
Jiaozhou Bay

178. Unnyul Kumsanpo, a wetland reserve south of the Taedong River in South Hwanghae Province, is on the northeastern shore of the Yellow Sea in what country?
North Korea

179. You can view the Akaishi Mountains from Nihondaira, a plateau region near the Kunozan Toshogu Shinto shrine on Suruga Bay on what Japanese island?
Honshu

180. Impressive karst formations and cave systems can be found in Gunung Mulu National Park in what country?
Malaysia

181. The Miankaleh Peninsula, which juts out into the Caspian Sea, is located in what Iranian Province?
Mazandaran Province

182. The Shapotou District is located in the Tengger District in what country home to the Taihangshan Gorge?
China

183. My Son, a set of Hindu temple ruins, was constructed by the Kings of Champa. This archaeological site is located in the Quang Nam Province in what country?
Vietnam

184. Deosai National Park is home to beautiful plains near Skardu in what administrative territory of Pakistan?
Gilgit-Baltistan

185. Lake Qaraoun is located at the southern end of the Beqaa Valley in what country?
Lebanon

186. Name the spit that separates the north and south basins of the Dead Sea in Jordan.
Lisan Peninsula

187. The Dujiangyan Irrigation System, constructed around 256 B.C., is on the Min River, in the Min Mountains of what Chinese province?
Sichuan Province

188. The Mergui Archipelago and Moscos Islands are located off the coast of the Tanintharyi Region of what country?
Myanmar

189. Lagonoy Gulf, which borders the Bicol Peninsula, is separated from the Philippine Sea by the Caramoan Peninsula, near Mount Mayon in what country?
Philippines

190. The Tsiv-Gombori Range is the site of Mariamjvari Nature Reserve in the Kakheti Region of what country?
Georgia

191. The Derawar Fort is in Bahawalpur in what country?
Pakistan

192. Lake Khasan is southeast of Posyet Bay in what country home to Rungrado Island, on the Taedong River?
North Korea

193. The Hose Mountains, located between the watersheds of the Balleh and Balui Rivers, is in what Malaysian state?
Sarawak

194. You can watch spectacular dragon boat races at Xixi National Wetland Park, near what city that is the capital of Zhejiang Province?
Hangzhou

195. Name the highest point in Iran's Mazandaran Province, the second highest in Iran behind Mount Damavand.
Alam-Kuh

196. Majnoon Island is located near the Majnoon Oil Field, which is situated close to what Iraqi city on the Shatt-al Arab River and is home to a well known sports complex?
Basra

197. Petropavl is a city on the Ishim River in the northern part of what country's border with Russia?
Kazakhstan

198. The Sumela Monastery is located in the Pontic Mountains, in Trabzon Province in what country where one can find the Erdemir Steel Plant in Karadeniz Eregli?
Turkey

199. Zoige Marsh, home to the highest marsh area in the world, borders the Min Mountains to the east and is known as the Songpan Grasslands. This wetland is on what plateau?
Tibetan Plateau

200. What island is home to the Baliem Valley and Yapen Island, situated in Cenderawasih Bay?
New Guinea

201. The Sakhir Desert is home to Areen Wildlife Park and what palace in Bahrain?
Al-Sakhir Palace (Sakhir Palace)

202. The Kachura Lakes are in the Karakoram Mountains of what country?
Pakistan

203. The Ashdod Sand Dunes are located on the coastal plain of what Eastern Mediterranean country?
Israel

204. The oligotrophic Gokyo Lakes are along the Dudh Kosi River in what Nepali national park bordering Tibet's Chomolungma Nature Reserve?

Sagarmatha National Park

205. The Shengsi Islands, whose largest island is Sijiao Island, is located at the mouth of the Yangtze River and is part of what larger archipelago?
Zhoushan Islands

206. Lake Tharthar is in the Sinjar Mountains in what country that signed a treaty with Iran in 1975 known as the Algiers Accord?
Iraq

207. Wilpattu National Park, home to natural, sandy water basins, is located west of Anuradhapura in what country?
Sri Lanka

208. The Izu-Ogasawara Trench, where the Pacific Plate is being subducted under the Philippine Sea Plate, is near the Izu Islands and what other archipelago whose largest island is Chichijima?
Bonin Islands (Ogasawara Islands)

209. The Nemegt Basin, in the northwestern Gobi Desert, is in what Mongolian Province whose capital is Dalanzadgad?
Omnogovi Province

210. Tarout Castle is located on Tarout Island, which was once part of the Dilmun Civilization. This island is connected to Qatif by two causeways in the Eastern Province of what country?

Saudi Arabia

211. The Jiwani Coastal Wetland, a designated Ramsar Wetland, borders the Dasht River and Gwadar Bay and is located in what country?
Pakistan

212. The fourth largest city in Lebanon is known for its vineyards and agriculture. Name this city, the capital of the Beqaa Governorate.
Zahlé

213. The Dong Phayayen-Khao Yai Forest Complex is an UNESCO World Heritage Site in what country?
Thailand

214. The Jayawijaya Mountains and the Sudirman Range make up the Maoke Mountains on what island?
New Guinea

215. The Kurram Valley is in the northwestern part of what country?
Pakistan

216. Borjomi Gorge is located between the Trialeti and Meskheti Ranges on what river in Georgia?
Kura River

217. The Cyclops Mountains are north of Lake Sentani, near Jayapura. Jayapura is on Yos Sudarso Bay and is the capital of what Indonesian province?
Papua

218. The Ebino Plateau, which borders Mount Karakuni, is situated near Kagoshima Bay, on the southern coast of what Japanese island?
Kyushu

219. The Stone Forest in China is a group of limestone formations near Kunming, the capital of what province?
Yunnan Province

220. The Datca Peninsula, which borders the Gulf of Gokova, is located in the Mugla Province of what country home to Ayvalik Islands Natural Park?
Turkey

221. Palu is at the mouth of the Palu River and is capital of Central Sulawesi Province in what country?
Indonesia

222. The Meratus Mountains, on Borneo, divides an Indonesian province into two parts. Name this province, whose capital is Banjarmasin.
South Kalimantan Province

223. The Anambas Islands are part of the Tudjuh Archipelago in what Indonesian province whose capital is Tanjung Pinang?
Riau Islands Province

224. What major city in China is the site of the Emin Minaret?
Turpan

225. The Maluku Islands, located on the Halmahera Plate in the Molucca Sea Collision Zone, are known by what other name?
Spice Islands

226. Mount Yonoha is the highest point on what island that is the fifth largest by area in Japan?
Okinawa Island

227. Name the highland plateau region considered one of the coldest parts of the Korean Peninsula.
Kaema Plateau

228. The Potohar Plateau borders the Kala Chitta Range to the north and Salt Range to the south in what country?
Pakistan

229. The Bada Valley, also known as the Napu Valley, is in Lore Lindu National Park in what country?
Central Sulawesi Province

230. What dam on the Balui River in the Malaysian state of Sarawak is the largest in Southeast Asia and is near Belaga?
Bakun Dam

231. Name the city on the Irrawaddy River that is considered Upper Burma's main economic and commercial center.
Mandalay

232. Shipton's Arch, also known as the Heavenly Gate, is one of the world's tallest natural arches and is northwest of Kashgar in what Chinese autonomous region?
Xinjiang Uyghur Autonomous Region (Xinjiang Uyghur or Xinjiang)

233. The Asmat Swamp, a wetland with mangroves and nipa palms, forms part of Lorentz National Park. This area, the largest in Southeast Asia, is in what country?
Indonesia

234. The Trango Towers are home to some of the largest cliffs in the world. These impressive rock formations can be found in the Baltoro Muztagh Mountains of Gilgit-Baltistan in what country?
Pakistan

235. The Mausoleum of Khoja Ahmed Yasawi is located in what Kazakh city near the Syr Darya River?
Turkestan

236. The oasis city of Serakhs is in the Ahal Region of what country home to the Kopet Dag Mountain Range?
Turkmenistan

237. Gosaikunda Lake in Langtang National Park is a religious site in what country containing the Dundwa Range?
Nepal

238. The Seven Star Crags are a series of rock formations near the sacred Buddhist site of Mount Dinghu, known as the "Green Gem on the Tropic of Cancer", and are in what Chinese province?
Guangdong Province

239. The Iran Mountains are situated on the border between Indonesia and what country with the Kelabit Highlands?
Malaysia

240. The Asagiri Plateau lies at the base of Mount Fuji in Fujinomiya. This plateau lies near the Isu Peninsula on what Japanese island?
Honshu

241. The Dong Phaya Yen Mountains extend from the Phetchabun Mountains to the northern side of the Sankamphaeng Range in what country?
Thailand

242. You can see the Keane Bridge stretching across the Surma River in Sylhet, a city in what country?

Bangladesh

243. Lake Hammar, part of the Hammar Marshes, is in what country that shares the Hawizeh Marshes on its border with Iran?
Iraq

244. Karambar Lake can be found in what country's Ishkoman Valley?
Pakistan

245. Norton Couloir is a gorge on the northern face of what mountain on the Nepal-China border?
Mount Everest

246. What Lebanese city once known as Heliopolis is east of the Litani River and home to a Temple of Bacchus?
Baalbek

247. The Türkmenbaşy Gulf borders the Bala-Ishem Salt Marshes and is in Hazar Nature Reserve in what country?
Turkmenistan

248. The Kosi River cuts through Chatra Gorge, near the Mahabharata Range in what country?
Nepal

249. Ratargul Swamp Forest is a freshwater forested wetland in what country containing the Somapura Mahavira World Heritage Site?

Bangladesh

250. The Tumen River, which originates on the slopes of Paektu Mountain, forms the border between Russia and what other country?
North Korea

251. What freshwater lake in Israel is also known as Lake Tiberias?
Sea of Galilee

252. Zhaoqing is located in what Chinese province with the Leizhou Peninsula in its southwestern region?
Guangdong Province

253. The Ma'in Hot Springs are in the Madaba Governorate of what country home to the Jabal Umm Fruth Bridge?
Jordan

254. Pedra Branca is the easternmost island in what country with the Kallang Basin?
Singapore

255. Pamukkale is a natural attraction with hot springs and bright pools formed by travertine terraces and carbonate minerals. This World Heritage Site is near the Kaklik Cave, known as the "Underground Pamukkale", in the Büyük Menderes River Valley in what country's Denazli Province?
Turkey

Europe

Includes European and Asian Russia

1. The Basilica of San Domenico is located in Siena in what Italian region home to the Medici Villas?
   Tuscany

2. The world's longest road tunnel is located northeast of Bergen in which Scandinavian country?
   Norway

3. Lake Constance lies on the border between Switzerland, Germany, and which other country?
   Austria

4. Kayakers can paddle on Lake Baikal in which country?
   Russia

5. The Ta' Cenc Cliffs can be found at the village of Sannat on Gozo Island in what country?
   Malta

6. Matka Canyon is home to Matka Lake and is located in what country whose capital is Skopje?
   Macedonia

7. The Yorkshire Dales, located in Yorkshire Dales National park in Northern England, are situated in what mountain range known as the "Backbone of England"?
    Pennines (Pennine Mountains)

8. The Douro River flows from its source in the Soria Province of Spain, and empties out into the Atlantic Ocean in Portugal at what city home to the Clerigos Church and the Palacio da Bolsa?
    Porto

9. St. Mortiz is an alpine resort and village in the Engadine Region, which follows the course of the Inn River from its source at the Maloja Pass in what country?
    Switzerland

10. The Ring of Kerry is east of the Iveragh Peninsula, which is home to MacGillycuddy's Reeks. This mountain range is home to Carrauntoohil and is in what country?
    Ireland

11. Delphi is a World Heritage Site in Greece near the Gulf of Corinth and overlooking the ancient region of Phocis. Delphi is on the southwestern slope of what mountain made of limestone?
    Mount Parnassus

12. Lamac Gate is a tectonic feature in the Devin Carpathians of what country home to the ruins of Devin Castle?
    Slovakia

13. The Rapperswil Peninsula juts out into what lake in Switzerland formed by the Linth River?
Lake Zurich

14. The Kocani Valley can be found along the Bregalnica River, the second largest river in what country?
Macedonia

15. The Trola Peninsula borders the Sado River in the Grandola Municipality of what country?
Portugal

16. The Dukla Pass is situated in the Laborec Highlands and is on the border between Slovakia and what other country?
Poland

17. The fortified city of Mdina is home to the Palazzo Santa Sofia, the Mdina Gate, and the Vilhena Palace in what archipelagic country?
Malta

18. The Tice Wetlands are located in the Latorica Protected Landscape Area in what country?
Slovakia

19. The highest point in the Sar Mountains is Titov Vrv, which is near Tetovo in what country?
Macedonia

20. Maninska Gorge is the narrowest canyon in what country where the Vihorlat Mountains can be found?
Slovakia

21. Cabo do Roca is located in Sintra-Cascais Natural Park in what country?
Portugal

22. The largest city in the Bergisches Land is in the Wupper Valley. Name this German city.
Wuppertal

23. The Saint-Gotthard Massif is home to the peak of Pizzo Rotondo in what country?
Switzerland

24. Ostrvica is a mountain in what country whose Rugova Canyon is located in the Prokletije Mountains?
Kosovo

25. Karkinit Bay borders the Isthmus of Perekop in what country?
Ukraine

26. Aggtelek Karst is a karst area in Aggtelek National Park, home to Baradla Cave in what country?
Slovakia

27. Vihren is the highest peak of a Bulgarian mountain range that is separated from the Rila Mountains by the Razlog Valley to the north. Name this mountain range.
Pirin Mountains

28. Bielefeld is located north of what forest in Germany bordering the Lippe Uplands to the east?
Teutoburg

29. The Boltysh Crater is located in the Kirovohrad Oblast of what country bordering the Black Sea?
Ukraine

30. Catania is on the eastern coast of Sicily, at the base of what volcano?
Mount Etna

31. Doiran Lake is shared by Greece and what other country?
Macedonia

32. In what city on the Guadalquivir River can you visit a Roman bridge and the Puerta del Puente?
Cordoba

33. The Blatnica Valley can be found in the Vel'ka Fatra Mountain Range in what country?
Slovakia

34. Name the upland region in Northern Ireland that contains the Glenshane Pass and has its highest point at Sawel Mountain.
Sperrin Mountains

35. Santa Barbara Castle is in Alicante and is situated on what mountain?
Mount Benacantil

36. Pfannenstiel is a mountain overlooking the Zurcher Oberland in what country?
Switzerland

37. The headland of Ras ir-Raheb borders Fomm ir-Rih Bay and is located near Bahrija in what country?
Malta

38. The Nördlinger Ries Crater is located near the Swabian Jura in what country?
Germany

39. The Dom Tower can be found in what city cut by the Oudegracht Canal?
Utrecht

40. The Sanok-Turka Mountains are located between the San and Stryi River Valleys on the border between Poland and what country?
Ukraine

41. Drenica is a region in the central part of what country that was once part of Yugoslavia?
Kosovo

42. Name the highest group of mountains in continental Portugal, which are also known as the Star Mountains and are home to the Vodafone Ski Resort.
Serra da Estrela

43. The Kozjak Hydro Power Plant is situated on the Treska River in what country?
Macedonia

44. The Iskar River cuts through Pancharevo Gorge near the Vitosha Mountains, and is the longest river in what country?
Bulgaria

45. The Murcia Cathedral is located in Murcia, a city in Spain on what river?
Segura River

46. The Odessa Catacombs are a group of tunnels under the port city of Odessa in what country?
Ukraine

47. The Masurian Lake District is home to more than two thousand lakes, including Lake Marmy. Lake Marmy is the second largest lake in what country?
Poland

48. The Bieszczady Mountains is located near the Lupkow Pass, which is situated on the border between Poland and what other country?
Slovakia

49. The Cotentin Peninsula extends out into the English Channel and borders the bay of Mont Saint-Michel. This landform is located in what country?
France

50. Name the historical capital of Banat, home to Dan Păltinişanu Stadium.
Timisoara

51. Name the mountain range in the western Alps that lies between Lake Geneva and the Mont Blanc Massif, and borders the Rhone Valley.
Chablais Alps

52. The Vasyugan Swamp is located in southwestern Siberia, along what river that forms at the confluence of the Biya and Katun Rivers?
Ob River

53. The Ondavska Highlands are located in the Presov Region of what country?
Slovakia

54. Selvagem Grande Island belongs to the Savage Islands, an archipelago in what country?
Portugal

55. The highest point in the Matra Mountains is also the highest point in Hungary. Name this landform.
Kékes

56. What major Central European river flows through Wroclaw in the Silesian Lowlands and empties out into Szczecin Lagoon?
Oder River

57. The world's largest inland harbor belongs to a German city near the confluence of the Rhine and Ruhr Rivers. Name this city.
Duisburg

58. What major Bulgarian seaport on the Black Sea is on a lake and a gulf of the same name?
Varna

59. Biskupin is an archaeological site in the Kuyavian-Pomeranian Voivodeship in what country?
Poland

60. Bauschanzli is an island and public park in what Swiss city whose Schauspielhaus Zurich theater is one of the most important in the German-speaking world?
Zurich

61. Name the third largest lake in Italy, near the Grigna Mountain Massif and the city of Lecco.
Lake Como

62. The Burghausen Castle is the longest castle in the world and is in what administrative district of Germany?
Upper Bavaria

63. One can find hoodoos at the Demoiselles Coiffees de Pontis rock formation near Lake Serre-Ponçon in what country?
France

64. The Hanko Peninsula is the southernmost point on the mainland of what Nordic country?
Finland

65. The Logoisk Crater is an impact crater in Belarus near what city in the Lahoysk Raion and the Minsk Region?
Lahoysk (Logoisk)

66. The Ruinaulta Canyon, a famous place for rafting, is accessible by the Rhaetian Railway in what country?
Switzerland

67. The Lincolnshire Wolds are a continuation of the Yorkshire Wolds and are in what political division of the United Kingdom?
England

68. The Sul'ov Rocks are part of a national nature reserve in the Sul'ov Mountains of what country?
Slovakia

69. The Mirusha Waterfalls are part of Mirusha Park, a regional park and protected area in what Balkan country?
Kosovo

70. Faro is the capital of what Portuguese region home to famous cliffs at Praia da Marinha Beach?
Algarve Region

71. El Pas de la Casa is a ski resort in the Encamp Parish of what country?
Andorra

72. Tuffeau stone is a type of limestone used to built the walls at the Chateau de Loches in what French valley?
Loire Valley

73. Gauli Glacier is in the canton of Berne in what country?
Switzerland

74. The Colle di Cadibona Pass is located in the Ligurian Alps in what country?
Italy

75. The Krkonose Mountains have their highest peak at Snezka, on the border between Poland and what country?
Czech Republic

76. What is the second largest city in Lesser Poland, home to a medieval castle near its center?
Lublin

77. Duf Falls is located in Mavrovo National Park near the Saint Jovan Bigorski Monastery in what country?
Macedonia

78. Kazantyp is a headland located on the Crimean Peninsula and bordering the Sea of Azov in what country?
Ukraine

79. Snowdonia is a mountainous region also known as Eryri in a language native to a region of the British Isles. Name this language.
Welsh

80. The Bay of Kotor is located in the Adriatic Sea and borders what country?
Montenegro

81. Name the only landmass in the Strait of Tartary that is situated off the southwestern coast of Sakhalin Island.
Moneron Island

82. The Snowcastle of Kemi is the world's largest snow fort, and is rebuilt every winter at Kemi, a city at the mouth of the Kemijoki River in what country?
Finland

83. You can view the Buskett Gardens and the president's summer residence, the Verdala Palace, from the city of Dingli in what country?
Malta

84. Lubusz Land is a cultural and historical area found around what major river in Poland, Germany, and the Czech Republic?
Oder River

85. Liberdade Square and Clerigos Church are historical sites in a city situated along the Douro River. Name this city.
Porto

86. Rusne Island lies in the Nemunas River Delta, at the entrance to the Baltic Sea in what country?
Lithuania

87. Plovdiv is on the banks of the Maritsa River, which originates in the Rila Mountains. Plovdiv is known as the "City of Seven Hills" and is the second largest city in what country?
Bulgaria

88. Gran Paradiso is situated between the Aosta Valley and what region of Italy whose capital is Turin?
Piedmont

89. The Nymphenburg Palace, which dates back to the Baroque Period, can be found in a city in the German state of Bavaria. Name this city.
Munich

90. The Radishchev Art Museum is located in the Russian city of Saratov, a major port city on what river?
Volga River

91. Morskie Oko is the deepest lake in the Tatra Mountains and translates to "Sea Eye" in a language native to what country containing its basin?
Poland

92. Mount Narodnaya is the highest peak in what mountain range?
Ural Mountains

93. The Iberdrola Tower is in what city in Spain located in the Basque Mountains?
Bilbao

94. Name the lake that is created from damming the Ibar River and is on the border between Kosovo and Serbia.
Gazivoda Lake

95. The Iron Gates is a gorge in Iron Gates National Park and Derdap National Park, on the border between Romania and what other country?
Serbia

96. The Zeta Plain extends to Lake Skadar, also known as Lake Scutari, to the south. This plain is also a lowland in what country?
Montenegro

97. Nottingam, a city surrounded by the Sherwood Forest, is situated on the River Trent in what country?
United Kingdom

98. Wolf Cave is located in Kristinestad, on the shore of the Bothnian Sea in what country?
Finland

99. Brescia is a city near Lake Garda and Lake Iseo in the northern part of what country?
Italy

100. The Azure Window is a limestone natural arch, near Fungus Rock on what Maltese island?
Gozo

101. The Annunciation Cathedral, the largest in Eastern Europe, is in what Russian city home to Freedom Square?
Kharkiv

102. What large city in Western Poland can be found in the Silesian Lowlands and along the Oder River?
Wroclaw

103. Name the tripoint at which the borders of Finland, Sweden, and Norway meet.
Three-Country Cairn

104. The Silfra Rift is located at a divergent plate boundary between the North American and Eurasian Plates in Thingvallavatn Lake in a national park in what country?
Iceland

105. The urban quarter of Euralille is in the French city of Lille, which is located on what river?
Dule River

106. The Leipzig Riverside Forest is a riparian forest in Leipzig, the largest city in what German state?
Saxony

107. Norsminde Fjord is located south of Aarhus on what Danish peninsula?
Jutland Peninsula

108. Name the largest wetland in Lithuania, within Dzukija National Park.
Cepkeliai Marsh

109. The Basodino Glacier is situated in the Lepontine Alps in the Italian-speaking Ticino Canton in what country?
Switzerland

110. Name the island that is the southernmost point of Malta, whose name may have come from the Arabic word for black pepper.
Filfla Island

111. Stromboli Island is located in the Tyrrhenian Sea and contains an active volcano. This Italian island is situated to the north of what larger island?
Sicily

112. The bight of Fyns Hoved separates the Great Belt from the Kattegat Strait and borders the island of Funen in what country?
Denmark

113. The Deliblato Sands are in Vojvodina Province, on the Pannonian Plain in what country?
Serbia

114. The De Long Islands are an archipelago northeast of the Anzhu Islands and belong to what country?
Russia

115. Name the highest point on the Massif des Calanques, a mountain to the southeast of Marseille.
Mont Puget

116. The longest Finnish place name has been given to a bog region in what region home to Riisitunturi National Park?
Lapland

117. The Trepca Mines are part of an industrial complex in Mitrovica in what Balkan country?
Kosovo

118. The tallest sand dune in Europe is located near Arcachon Bay and Cap Ferret. Name this landform.
Dune of Pilat (Great Dune of Pyla)

119. Kolindsund is a drained lake on the Djursland Peninsula in what country home to Mols Bjerge National Park?
Denmark

120. The Beskids are a series of mountain ranges that span four countries and are part of what larger Central European range?
Carpathian Mountains

121. The Moravian-Silesian Beskids are a range of mountains in the Czech Republic, and extend a little into what neighboring country?
Slovakia

122. The Susa Valley borders the Cottian Alps to the south in France and what other country?
Italy

123. Tornetrask is a lake near Abisko National Park and is the source of the Torne River in what country?
Sweden

124. Kiaules Nugara is an island at the Port of Klaipeda in what Baltic country?
Lithuania

125. The Blędów Desert is a sandy area on the Silesian Upland in what country?
Poland

126. The Fehmarn Belt, a strait connecting the Bay of Mecklenburg and the Bay of Kiel, borders what Danish island to the north?
Lolland

127. What is the name of the highest peak in Belarus, located near the cities of Minsk and Dzyarzhynsk?
Dzyarzhynskaya Hara

128. The Upper Barrakka Gardens are a public garden from which you can view the Grand Harbor and the Marsamxett Harbor of what country?
Malta

129. The Lurgrotte Karst Cave is the largest cave in what mountain range of Austria?
Eastern Alps

130. Name the landform that separates the Curonian Lagoon from the Baltic Sea in Lithuania and the Kaliningrad Oblast of Russia.
Curonian Spit

131. The Vechtstreek region is located on both sides of the Vecht River on the border between two provinces belonging to what country?
Netherlands

132. Cima d'Asta is the highest peak in the Fiemme Mountains in what country?
Italy

133. The Oura Archipelago is part of Bothnian Sea National Park near what Finnish city on the Kokemaenjoki Estuary?
Pori

134. Guldborgsund is a strait bordering the island of Lolland and what other island where the living history museum of Middelaldercentret can be found?
Falster

135. What city on the Warta River is near the Morasko Meteorite Nature Reserve in Poland?
Poznan

136. The Landes Forest is the largest maritime-pine forest in Europe, and is in what country's Aquitaine region?
France

137. The Rabjerg Mile is a moving sand dune near what Danish town that is the northernmost settlement in Denmark and is located on the Skagen Odde Peninsula?
Skagen

138. Name the low mountain range located in the Rhenish Massif on the border between Belgium and Germany and whose highest point is Hohe Acht.
Eifel Mountains

139. The Golden Stag International Music Festival is held annually in Brasov, a city home to the Poiana Brasov Ski Resort in what country?
Romania

140. Ben Nevis, the highest mountain in the United Kingdom is located in what mountain range?
Grampian Mountains

141. The Aarhus River can be found flowing through Brabrand Lake in what country?
Denmark

142. The world's tallest manmade waterfall, the Cascata delle Marmore, is located near Terni, a city on the Nera River Plain in what country?

Italy

143. Sognefjorden is the second largest fjord in the world, and extends for more than 120 miles. This fjord, along with Naeroyfjord, is a famous area in what country?
Norway

144. The mountain cliffs of Slieve League in County Donegal are nearly three times the height of the Cliffs of Moher in County Claire. These landforms are in what country?
Ireland

145. The Timok River flows into the Danube River near Bergovo, a town in what country whose Belasica Mountains are shared with Macedonia and Greece?
Bulgaria

146. Lille Vildmose is a raised bog which at one time extended to the Rold Skov Forest in what country?
Denmark

147. Lagoa das Sete Cidades is a twin lake on Sao Miguel Island in what country?
Portugal

148. The Mourne Wall can be found in the Mourne Mountains, a range in what political division of the United Kingdom?
Northern Ireland

149. The Metohija Basin is located in the western part of what landlocked Western Balkan country?
Kosovo

150. What peninsula of Denmark is also known as the Cimbrian Peninsula and is separated into three areas in its Danish region?
Jutland Peninsula

151. Name the second largest city in the Netherlands' Utrecht Province, home to the medieval gate of Koppelpoort.
Amersfoort

152. The peninsula of Odsherred is located on what Danish island that is the most populous in the country and connected to Funen Island by the Great Belt Bridge?
Zealand (Sjaelland)

153. Name the largest artificial lake in Belarus, located in the country's Minsk Region.
Vileyka Reservoir

154. The Ggantija Megalithic Temple Complex and Xaghra Stone Circle are Neolithic sites in what country?
Malta

155. The Valli di Comacchio lagoons are located in the Emilia-Romagna Region of what country?
Italy

156. Lake Narach can be found in the Neris River Basin and is the largest lake in what country?
Belarus

157. The Sonian Forest is located near what major Belgian city home to the Parc du Cinquantenaire?
Brussels

158. Lake Keitele and Lake Kolima are located in the central part of what country?
Finland

159. The Cranborne Chase is a chalk plateau many miles southwest of the Chiltern Hills in what country?
United Kingdom

160. Name the wetland region also known as the Pripet Marshes or the Rokitno Marshes, which surround the Pripyat River in Belarus.
Pinsk Marshes

161. What dam is located in southern Portugal, on the Guadiana River, and created one of the largest artificial lakes in Western Europe?
Alqueva Dam

162. Visby is a city and UNESCO World Heritage Site on what Swedish island?
Gotland

163. Name the major river island in the Sozh River of Belarus that shares its name with a nearby city.
Vetka Island

164. Tigne Point is a peninsula in the town of Sliema in what country?
Malta

165. The Saltfjellet Mountain Range is located in the cultural area of Sapmi, also known as Lapland, in what country?
Norway

166. Lahemaa National Park is located in what Baltic country?
Estonia

167. Ninu's Cave and Xerri's Grotto are caves in what country whose capital is Valletta?
Malta

168. Dunes cover the landscape of Curonian Spit National Park, one of the five national parks of what country?
Lithuania

169. The High Fens is a plateau region in the Liège Province of what country?
Belgium

170. The Camargue is a recognized wetland as well as the largest river delta in Western Europe. This region is the

site where what major river that originates at Lake Geneva flows into the Mediterranean Sea?
Rhone River

171. The Giant's Causeway connects tens of thousands of basalt columns in County Antrim in what country?
United Kingdom

172. Although the Netherlands is a country with much of its land below sea level, its highest point is a famous hill in the country's Limburg Province. Name this landform.
Vaalserberg

173. You can find vineyards in the Montferrat, a region and World Heritage Site divided by the Tanaro River in what country?
Italy

174. Olando kepure, also known as Dutchman's Cap, is a hill in Seaside Regional Park, on the Baltic Sea coast in what country?
Lithuania

175. Kiskunsag National Park is located in the Danube-Tisza Interfluve, in the Pannonian Basin in what country?
Hungary

176. Matsalu National Park, on the Kasari River delta, is located next to the East Atlantic Flyway in what country?
Estonia

177. Hammerso is the largest lake on what Danish island southwest of the Ertholmene Archipelago and whose highest point is Rytterknaegten?
Bornholm

178. Name the lake that is the largest by area in Sweden and the third largest by area entirely in Europe.
Lake Vanern

179. The Ŝar Mountains extend into the tripoint border of three countries including Macedonia, Albania, and what other country?
Kosovo

180. The Cenotaph War Memorial is made up of Portland stone and is on Whitehall Road in what major British city?
London (The City of Westminster is a borough, not a city)

Africa

1. The Verneukpan Salt Flat is approximately eighty miles away from Upington, a city on the Orange River and near Augrabies Falls in what country?
   South Africa

2. The peak of Tullu Dumtu is on the Sanetti Plateau in Bale Mountains National Park in what country?
   Ethiopia

3. The Rwenzori Mountains, sometimes called the "Mountains of the Moon", are home to Virunga National Park. This range is on the border between Uganda and what other country?
   Democratic Republic of the Congo

4. Bamako, the capital of Mali, is located on what river whose source is in the Guinea Highlands?
   Niger River

5. Tugela Falls is located near the border between Lesotho and which other country?
   South Africa

6. What lake, also known as Dembiya, is the source of the Blue Nile and is the largest lake by area in Ethiopia?
   Lake Tana

7. The Aberdare Range whose highest peak is Mount Satima, is located in the Nyandarua County of what Kenyan city?
Nairobi

8. The Dawa and Ganale Dorya Rivers meet and form the Jubba River on Ethiopia's border with what country situated below the Gulf of Aden?
Somalia

9. The Orange River originates in the Maloti Mountains, a subrange of the Drakensberg Mountains, in what country?
Lesotho

10. Sibiloi National Park lies on the eastern shore of what major East African lake home to Central Island, a volcanic island that is the location of Central Island National Park?
Lake Turkana

11. The Semien Mountains are northeast of Gondar, a city located on the Lesser Angereb River. This mountain range is in what country?
Ethiopia

12. The Ankify Peninsula, in the Diana Region, is near the islands of Nosy Be and Nosy Komba in what country home to Lake Tsimanampetsotsa?
Madagascar

13. The Gariep Dam, situated on the Orange River, is located in the central part of what country with the provinces of KwaZulu-Natal and the Eastern Cape?
South Africa

14. Keran National Park is located in the northern region of what country in West Africa whose capital city, Lome, is in the Golfe Prefecture?
Togo

15. The Richat Structure, known as the "Eye of the Sahara", is a prominent circular feature in what country whose second largest city, Nouadhibou, is situated on the Ras Nouadhibou Peninsula?
Mauritania

16. Mgahinga Gorilla National Park, in the Virunga Mountains, is contiguous with Rwanda's Volcanoes National Park and is in what country?
Uganda

17. Ramciel, sometimes considered the geographic center of South Sudan, will serve as the future capital of the country and is located in what South Sudanese state?
Lakes State

18. The Hoanib River, located a few miles north of Skeleton Coast Park, is in what country bordering Botswana?
Namibia

19. Lion's Head is between Table Mountain and Signal Hill in Table Mountain National Park in what South African city?
Cape Town

20. Le Morne Brabant is a peninsula and UNESCO World Heritage Site in what country home to Aapravasi Ghat?
Mauritius

21. The Itzehi-Tezhi Dam is on the Kafue River in what country home to the Liuwa Plain?
Zambia

22. Lake Manyara National Park is located on Lake Manyara, which is east of what major Tanzanian lake?
Lake Eyasi

23. Lake Ngami, south of the Okavango Delta, is in what country home to Central Kalahari Game Reserve?
Botswana

24. Augrabies Falls is on the Orange River, which forms part of the border between South Africa and what country?
Namibia

25. Bombetoka Bay is a body of water that can be found near Mahajanga in what island country?
Madagascar

26. The Cosmoledo Group and the Aldabra Islands are part of what country?
Seychelles

27. The Erg of Bilma is a sea of sand dunes in what desert region of Niger?
Tenere

28. Dzanga-Ndoki National Park is located near Berberati, the third largest city in what country?
Central African Republic

29. Addis Ababa Bole International Airport and the headquarters of the African Union are located in Addis Ababa in what northeast African country?
Ethiopia

30. The major city of Dese is approximately fifty miles north of the peak of Abuye Meda in what landlocked country?
Ethiopia

31. Borgu is a region in the Niger State of Nigeria and the northern region of what country home to Djougou?
Benin

32. The Yoruba people make up the majority of the population of Ogbomosho, in what Nigerian state home to Old Oyo National Park and whose capital is Ibadan?
Oyo State

33. Lake Kossou, an artificial lake on the Bandama River, caused the displacement of 85,000 Baoule people and is the largest lake in the central region of what French-speaking country?
Cote d'Ivoire (Ivory Coast)

34. Orango Islands National Park is home to a rare species of saltwater hippo, and is located off the coast of what country bordering Guinea to the east?
Guinea-Bissau

35. The Kharga Oasis and the Dakhla Oasis are located in the south central part of what country whose largest peninsula is located in Asia?
Egypt

36. Sprawling orange sand dunes make up the impressive terrain of the Lompoul Desert. You can find his desert about ninety miles south of Saint-Louis in what country?
Senegal

37. Cap d'Arguin is a cape jutting out into Levrier Bay in what country?
Mauritania

38. Taulud Island is a part of the Dahlak Archipelago in what country bordering the Red Sea?
Eritrea

39. Western Sahara's smallest border is with what country home to the natural region of Tademait?
Algeria

40. Nzerekore is west of the Nimba Mountains in what West African country that borders the Atlantic Ocean?
Guinea

41. The city of Marsabit, almost completely surrounded by Marsabit National Park, lies southeast of the Chalbi Desert in what Northeast African country?
Kenya

42. The Okavango River, in Namibia and Botswana, drains into the Moremi Game Reserve and fills Lake Ngami. Also known as the Cubango River, it flows through what other country?
Angola

43. The Bou Regreg is a river separating the commuter town and city of Sale from what other city on the opposite bank home to the Hassan Tower?
Rabat

44. Hwange National Park is in the Matabeleland Province and is the largest national park by area in what country?
Zimbabwe

45. The Lekki Peninsula is situated east of what major coastal city in Nigeria that is located in a state of the same name?
Lagos

46. Sekondi and Tema are cities near the coast of what country home to Lake Volta, one of the world's largest artificial lakes and Kakum National Park?
Ghana

47. Lake Assal is located west of Day Forest National Park in what country?
Djibouti

48. Beledweyne is a major city on the Shebele River, which flows into the Jubba River near Jamaame in what country?
Somalia

49. Kisumu is a major port city on Lake Victoria. This city is in the western part of what country in the Great Rift Valley?
Kenya

50. Kitgum, a city on the Pager River, is located in the northern region of what country where the Kazinga Channel separates Lake George from Lake Edward?
Uganda

51. Bandundu is located on the eastern bank of the Kwango River in what country bordering Cabinda?

Democratic Republic of the Congo

52. What city in Benin is home to the Ancien Pont Bridge and the economically important Dantokpa Market?
Cotonou

53. Conkouati-Douli National Park is located miles northwest of the cities of Dolisie and Pointe-Noire in what country?
Central African Republic

54. The Lagdo Reservoir is located southeast of the major city of Garoua in what country?
Cameroon

55. Pongara Point, which juts out into the Gulf of Guinea, is located on a piece of land on one side of the city of Libreville in what country?
Gabon

56. The Bight of Bonny forms the northern coast of what island that is the second largest piece of land in Equatorial Guinea?
Bioko

57. Lake Chilwa borders Mozambique and what other country home to the Shire River and the city of Blantyre?
Malawi

58. The Quirimbas Archipelago belongs to what country that is bisected by the Zambezi River and is the site of the Zambezi River Delta?
Mozambique

59. The Matopos Hills are about fifty miles south of Bulawayo, one of the most populous cities in what country where Mutare is located on the Save River?
Zimbabwe

60. Onitsha and Idah are cities along the Niger River in what country bordering the Bight of Benin?
Nigeria

61. Travertine hot springs make up the thermal wonder of Hammam Maskhoutine, home to multicolored walls of calcium carbonate and iron. This site is in the Guelma Province of what country?
Algeria

62. The Mlilwane Wildlife Sanctuary is located in the Ezulwini Valley near the Mdzimba Mountains. This wildlife sanctuary is located in what southern African country?
Swaziland

63. What major Somalian city, located in the Ogo Mountains, is situated near Gabiley, a well-known agricultural center in the Woqooyi Galbeed Administrative Region?
Hargeisa

64. Mount Nyiru is located southwest of what major lake in Kenya that was formerly known as Lake Rudolf and is the world's largest permanent desert lake?
Lake Turkana

65. The Itombwe Mountains, located in the South Kivu Province, are located on the western shore of Lake Tanganyika in what country?
Democratic Republic of the Congo

66. The Lamadaya Waterfalls are in the Cal Madow Mountain Range. This range, whose highest peak is Mount Shimbiris, is in the northeastern part of what country that has had a famous history of pirates?
Somalia

67. The Karre Mountains, whose southern elevations form part of the watershed of the Congo River, are home to the peak of Mount Ngaoui. Mount Ngaoui is the highest peak in what country?
Central African Republic

68. The Mousa Ali Volcano is at the tripoint border of three Northeast African countries. This stratovolcano, in the Tadjourah Region, Southern Red Sea Region, and Afar Region, is the highest point in what country?
Djibouti

69. Piton de la Petite Riviere Noire is the highest point in what island country located in the Mascarene Islands in the Indian Ocean?
Mauritius

70. Mutsamudu is the capital of what island in Comoros that borders the Mozambique Channel to the southwest?
Anjouan

71. Desroches Island is the main island of the Amirantes Islands, an archipelago in the Outer Islands of what country whose official currency is the rupee?
Seychelles

72. Kunta Kinteh Island is a river island and a World Heritage Site. It is in the western part of the Gambia River in what country?
Gambia

73. What Guinean city is on the Kaloum Peninsula and is home to the Sandervalia National Museum and a St. Mary's Cathedral?
Conakry

74. Ziguinchor, the capital of the Ziguinchor Region, is at the mouth of the Casamance River in what country?
Senegal

75. Lake Alaotra is the source of the Maningory River and is located in the Toamasina Province. This lake, along with the Bemolanga Oil Fields, are located in what country?
Madagascar

76. The Kagera River, which originates in Lake Rweru in Burundi, flows to a confluence with the Ruvubu River near Rusumo Falls on the border between Tanzania and what other country?
Rwanda

77. The Azawagh is a dry basin made up of flatlands in Niger, Algeria, and what West African country?
Mali

78. Mount Kinyeti is the highest peak in the Imatong Mountains, which are found in Imatong State. This peak, near the border with Uganda, is the highest point in what country?
South Sudan

79. A part of the Goda Mountains are protected within Day Forest National Park. This mountain range is home to what nation's most heavily vegetated area?
Djibouti

80. Control over Rio Muni in present day Equatorial Guinea passed from Portugal to what other Western European country in 1778 by the Treaty of El Pardo?
Spain

81. The Succulent Karoo is a desert region of South Africa and what other country home to Namib-Naukluft National Park, which encompasses the Naukluft Mountains and the Namib Desert?
Namibia

82. Lake Tshangalele is an artificial lake created by a dam, located on the Katanga Plateau. This plateau, which is home to the source of the Lufira River, is located in what country?
Democratic Republic of the Congo

83. Sossusvlei is a salt and clay pan, as well as an endorheic drainage basin, and is home to the Sesriem Canyon in what country?
Namibia

84. Phophonyane Falls Nature Reserve is located near the town of Piggs Peak. This nature reserve, along with Hlane Royal National Park, are large protected areas in what country?
Swaziland

85. The Nafusa Mountains are a major mountain range located in the western part of the Tripolitania Region of what country?
Libya

86. Ichkeul Lake is located near the city of Bizerte, the northernmost city on the continent of Africa. This lake and city are both located in what country?
Tunisia

87. The Ennedi Plateau and the Ouaddai Highlands are located in the eastern part of what country bordering Sudan?
Chad

88. The Ubangi River, which is formed at the confluence of the Mbomou and Uele Rivers, forms most of the border between the Democratic Republic of the Congo and what other country home to the Karre Mountains?
Central African Republic

89. The Tumba-Ngiri-Maindombe Wetland is the largest Ramsar Convention Wetland in the world. This wetland, more than twice the size of Maryland or Belgium, is located near Mbandaka in what country?
Democratic Republic of the Congo

90. The Vredefort Crater is located on the Witwatersrand Plateau, spanning parts of the Gauteng, North West, and Mpumalanga Provinces of what country?
South Africa

91. The Gulf of Zula is situated to the west of the Buri Peninsula in what small northeast African country?
Eritrea

92. Sehlabathebe National Park is located in the Maloti Mountains, a subrange of the Drakensburg Mountains, in what country?
Lesotho

93. The Caves of Hercules are located west of Cap Spartel, which is at the northern end of the Rif Mountains in what country?
Morocco

94. Comoe National Park is a biosphere reserve in the Zanzan and Savanes Districts. This park, the largest protected area in West Africa, is in the northeastern part of what country?
Cote d'Ivoire (Ivory Coast)

95. Mont Idoukal-n-Taghes is in the south-central part of the Air Mountains and at the northern end of the Bagzane Plateau. This mountain is the highest peak in what country?
Niger

96. Deadvlei is a white clay pan whose name has English and Afrikaans roots, and is a salt pan in what country?
Namibia

97. Tai National Park is home to one of the last remaining areas of primary rainforest in West Africa. This national

park, between the Cavally and Sassandra Rivers, can be found in what country?
Cote d'Ivoire (Ivory Coast)

98. The Atewa Range forms the southwest border of Lake Volta and is situated near the Kwahu Plateau in what country?
Ghana

99. Kongou Falls, reputed as one of the most beautiful waterfalls in Central Africa, is located within Ivindo National Park in what country?
Gabon

100. The Barotse Floodplain, which is located in the Western Province, spreads over part of former Barotseland. This floodplain, a famous wetland in southern Africa, is located on the Zambezi River in what country?
Zambia

101. Nyika National Park is on the Nyika Plateau, bordering the Luangwa Valley. This plateau, crossed by the South Rukuru River, is in Zambia and what other country?
Malawi

102. Lake Magadi is known for its population of flamingos and is north of Lake Natron in what country?
Kenya

103. Fomboni is the capital of the Federal State of Moheli and the third largest city in what island country home to Mount Karthala and Prince Said Ibrahim International Airport?
Comoros

104. The botanical garden of Jardin de l'Etat is in what territory of France?
Reunion

105. Pico Cao Grande is a volcanic plug in Obo Natural Park in what country home to the Pestana Equador Beach Resort on the island of Ilheu Das Rolas?
Sao Tome and Principe

106. The Mambilla Plateau, home to Gashaka Gumti National Park and Ngel Nyaki Forest Reserve, is located in the Taraba State of what country?
Nigeria

107. The Goualougo Triangle is located at the southern end of Nouabale-Ndoki National Park in what Central African country home to Lake Tele?
Republic of the Congo

108. The Toshka Lakes, which are recently formed endorheic lakes in the Sahara Desert, are located west of Lake Nasser in what country bordering Abu Qir Bay?
Egypt

109. The Cederburg Mountains are home to a nature reserve near the city of Clanwilliam in what country?
South Africa

110. The Nuba Mountains are in the southeast part of the Kurdufan Region, whose capital is Al-Ubbayid. This region is located in the center of what country?
Sudan

111. Mau Forest is the largest indigenous montane forest in East Africa, in what country?
Kenya

112. Piton des Neiges is the highest point in what French overseas territory?
Reunion

113. Ol Doinyo Lengai, the world's only active volcano that produces natrocarbonatite lava, is situated in the Gregory Rift in the Arusha Region of what country?
Tanzania

114. The Seven Colored Earths are a geological formation in the Riviere Noire District of what country home to the Vallee de Ferney?
Mauritius

115. Twyfelfontein is a World Heritage Site home to ancient rock engravings in what country's Kunene Region?
Namibia

116. The Avenue of the Baobabs is a line of baobab trees on both sides of a dirt road in the Menabe Region of what country?
Madagascar

117. Idjwi Island, the second largest inland island in Africa, is located in Lake Kivu and belongs to what country?
Democratic Republic of the Congo

118. Tadrart Acacus, a mountain range home to prehistoric rock art, colorful dunes, and wadis, is a World Heritage Site east of the city of Ghat in what country?
Libya

119. Lake Langano, on the border between the East Shewa and Arsi Zones, is in the Oromia Region of what landlocked country?
Ethiopia

120. The Kirstenbosch National Botanical Garden and the limestone caves of Sterkfontein are in what country home to uShaka Marine World and Gold Reef City?
South Africa

121. Kalandula Falls, located on the Lucala River, are located at the Black Rocks at Pungo Adungo in what country?
Angola

122. What island country, known as "Red Island", is home to Masoala National Park, mostly in the Sava Region?
Madagascar

123. Monrovia, located on Cape Mesurado, is the capital city of what West African country?
Liberia

124. The Boura Mountains and the Mabla Mountains are located in what country home to the Ras Siyyan Peninsula?
Djibouti

125. The Siwa Oasis, located between the Qattara Depression and the Great Sand Sea, was once part of Ancient Libya and is in the Matruh Governorate of what country?
Egypt

126. Lake Bangweulu, which is drained by the Luapula River, is located in the Luapula Province and Northern Province of what country home to Victoria Falls?
Zambia

127. Jebel Barkal Mountain and the historical city of Napata are located in the Nubia Region and are together a World Heritage Site in what country?
Sudan

128. The Quirimbas Islands are near Pemba, the capital of the Cabo Delgado Province in what country?

Mozambique

129. The Nyungwe Forest and Nyungwe Forest National Park are in the southwestern part of Rwanda at the border with the Democratic Republic of the Congo, Lake Kivu, and what country?
Burundi

130. The Cradle of Humankind is a paleoanthropological site and World Heritage Site near the Rising Star Caves in the Gauteng Province of what country?
South Africa

131. Douala, the capital of Littoral Region, is the largest city in what country with the Western High Plateau and Oku Volcanic Fields?
Cameroon

132. Antongil Bay is a body of water located in what country with many people of Indonesian descent?
Madagascar

133. What city in Ethiopia lies at the foot of Mount Entoto and near the Koka Reservoir?
Addis Ababa

134. Trou de Fer Canyon and Mafate Caldera are landforms in what territory in the Indian Ocean?
Reunion

135. The Aberdare Mountains, whose highest point is Mount Satima, falls steeply to the Kinangop Plateau and is located near Lake Naivasha in what country?
Kenya

136. The Obudu Plateau is on the Oshie Ridge of the Sankwala Mountain Range. This landform is a feature of the Cross River State in what country?
Nigeria

137. Lac de Mal is located in the Brakna Region of what country containing Banc d'Arguin National Park?
Mauritania

138. The Three Dikgosi Monument is a bronze sculpture in what country where the Tuli Block can be found?
Botswana

139. Ebrie Lagoon is located in what country?
Cote d'Ivoire (Ivory Coast)

140. Koutammakou, the Land of the Batammariba, is a World Heritage Site home to mud houses in what country?
Togo

141. Felicite Island, a forested granitic island, is east of La Digue in what country?
Seychelles

142. Lalla Khedidja is the highest summit of the Djurdjura Mountain Range, a subrange of the Atlas Mountains in what country?
Algeria

143. San Carlos is a shield volcano on the island of Bioko and is the second highest peak in what country?
Equatorial Guinea

144. The Pinselly Classified Forest is in the Mamou Prefecture, near Oure-Kaba in what country?
Ghana

145. Abokouamekro Game Reserve is located in what country sharing the Komoe River with Burkina Faso?
Cote d'Ivoire (Ivory Coast)

146. Kumasi, known as the "Garden City", was the former commercial, industrial, and cultural capital of the Ashanti Empire in what present day country?
Ghana

147. Mindelo, the largest city on Sao Vicente Island, is on Porto Grande Bay in what country home to the volcano of Pico do Fogo?
Cape Verde

148. The Balancing Rocks, geomorphological features of igneous rocks, are located in Matopos National Park in what country?

Zimbabwe

149. The Anjavavy Forest and the Mikea Forest are located in what country home to the Royal Hill of Ambohimanga?
Madagascar

150. Fazao Malfakassa National Park, between the Kara Region and Centrale Region, is the largest national park in what country?
Togo

151. Morne Seychellois is on the island of Mahe in a national park of the same name in what country?
Seychelles

152. Kakamega Forest is the only tropical forest in what country home to the Kit-Miyaki Rock Formation?
Kenya

153. Mont Abourassein is located in Haute-Kotto Prefecture of the Central African Republic and the Western Bahr el Ghazal State of what country?
South Sudan

154. The Nabemba Tower, also known as the Elf Tower, is named after Mont Nabemba. It is located in the Sangha Department, and is the highest point in what country?
Republic of the Congo

155. The Jwaneng Diamond Mine, the richest in the world, is located in what landlocked country?
Botswana

156. The Loriu Plateau, west of Barrier Volcano, borders the Suguta Valley to the east and is located in what country where the Ngong Hills can be found?
Kenya

157. Lake Sonfon and the Turner's Peninsula are located in what country home to the Banana Islands?
Sierra Leone

158. Chinguetti, a former medieval trading center, is home to the Friday Mosque of Chinguetti, and is a World Heritage Site in the Adrar Plateau Region of what country?
Mauritania

159. The Sa'ad ad-Din Islands, an archipelago situated near the ancient port city of Zeila, is administered by what de facto independent state?
Somaliland

160. Lake Abaya, whose southern shores are part of Nechisar National Park, sometimes overflows into Lake Chamo. These two lakes are located in what landlocked country whose Awash River is an UNESCO World Heritage Site?
Ethiopia

161. Cap Zebib and Cap Bon are in what country home to Zembra Island, a tall rock formation with a delicate ecosystem?
Tunisia

162. Pointe des Almadies, lies near Iles des Madeleines National Park and Leopold Sedar Senghor International Airport. Located on the Cap Vert Peninsula, this point is Africa's westernmost, and is in what country?
Senegal

163. The Aso Rock, Olumo Rock, and Zuma Rock are located in what country home to the Hadejia-Nguru Wetlands in the Yobe State?
Nigeria

164. What city in the Democratic Republic of the Congo is the capital of the Kasai-Oriental Province, and is situated on the Sankuru River?
Mbuji-Mayi

165. The deserts of Wadi Al-Hitan and Wadi El Natrun are located in the Al Fayyum Governorate and Beheira Governorate of what country?
Egypt

166. The seaport of Quelimane is the administrative capital of the Zambezia Province and the province's largest city. This city, whose local language is Chuwabo, is in what country?

Mozambique

167. The Fish River Canyon, the largest canyon in Africa, is part of Richtersveld Transfrontier Park and is a popular tourist attraction in what country home to the Diamond Coast?
Namibia

168. Sedudu, a fluvial island located on the Cuando River, is adjacent to the border with Namibia and is located in what country known for its many salt flats and salt pans?
Botswana

169. The Upemba Depression, home to Lake Upemba and Lake Kisale, is partially within Upemba National Park in the Katanga Province of what country?
Democratic Republic of the Congo

170. The Springbok Flats and the Koue Bokkeveld Mountain Range are located in what country?
South Africa

171. The Arrei Mountains, known as the "Roof of Ali Sabieh", are in what country home to Ardoukoba Volcano and the Seven Brothers Islands in the Bab Iskender Strait?
Djibouti

172. In what country can you find the Farafra Depression and the Bahariya Oasis?
Egypt

173. The Sikh Temple Makindu is a gurudwara in what Kenyan city known as the "Green City in the Sun" and is home to the world's only protected area found in a major city?
Nairobi

174. The Ben Youssef Mosque and the Bab Agnaou Gate can be found in Marrakech in what country?
Morocco

175. The iSimangaliso Wetland and uKhahlamba-Drakensberg Park are located in what province of South Africa bordering Swaziland, Lesotho, and Mozambique?
KwaZulu-Natal Province

176. The Ketchaoua Mosque combines Byzantine and Moorish architecture and is a World Heritage Site near the National Library and Great Mosque of what country?
Algeria

177. Cape Coast Castle and Elmina Castle are two of about forty large commercial forts and former slave castles built along the Gold Coast of Africa in what country?
Ghana

178. The Karisoke Research Center, between Mount Karisimbi and Mount Bisoke near the Congolese border, is located within Volcanoes National Park in what country?
Rwanda

179. Mkomazi National Park, which is contiguous with Kenya's Tsavo West National Park, is located in the Kilimanjaro and Tanga Regions of what country?
Tanzania

180. Hippopotamuses, which populate the region in and around Lake Tengrela, can be found in Mare aux Hippopotames National Park, the only biosphere reserve in what country?
Burkina Faso

181. The ancient archaeological site of Adulis, situated approximately thirty miles south of Massawa on the Gulf of Zula, was a port of the Kingdom of Aksum. This site is located in what country bordering the Red Sea?
Eritrea

182. Lake Amaramba, on the Nyasa Plateau, and the Bvumba Mountains, referred to as the "Mountains of the Mist", are geographical features of what country?
Mozambique

183. Cross River National Park is in the Cross River State of what country?
Nigeria

184. Bunce Island, on the Sierra Leone River, is in the harbor of Freetown in what West African country?
Sierra Leone

185. Jemaa el Fna is a famous square and marketplace in what Moroccan city near the Menara Gardens?
Marrakech

186. You can view Cape buffalo and Nile crocodiles at Mana Pools National Park in what country home to the Honde Valley?
Zimbabwe

187. Fasil Ghebbi, now an UNESCO World Heritage Site, was formerly the fortress of the emperors of what country home to the city of Bahir Dar?
Ethiopia

188. The ruins of Khami are near Bulawayo. This pre-colonial city, once the capital of the Butua Kingdom of the Torwa Dynasty, is in what country with Chapungu Sculpture Park and the National Heroes Acre Monument?
Zimbabwe

189. The Rubaga Cathedral, located on Lubaga Hill, is located in what major city of Uganda home to the Nakasero Market and Makerere University?
Kampala

190. The Draa Valley, home to the Fezouata Formations, is located in what country?
Morocco

191. The Panorama Route is a scenic road in the Mpumalanga Province of what country?
South Africa

192. The Morico and Crocodile Rivers are tributaries of what river that flows through four Southern African countries?
Limpopo River

193. Mount Mulanje, which is part of the protected Mulanje Mountain Forest Reserve, is a popular tourist attraction in what country?
Malawi

194. The Temple of Kom Ombo is a double temple which was constructed during the Ptolemaic Dynasty. This temple is located in the Aswan Governorate in what country?
Egypt

195. You can see thousands of flamingoes on the shores of Lake Nakuru in Lake Nakuru National Park. This soda lake and national park in the Great Rift Valley, which also protects Eastern Black Rhinos, is in what country?
Kenya

196. Bo-Kaap, formerly known as the Malay Quarter, is famous for its brightly colored houses, cobblestoned streets, and many cultures. It is on the slopes of Signal Hill in what country?
South Africa

197. The breathtaking view of Rhumsiki Rock in the Mandara Mountains is created by volcanic plugs and basalt outcroppings. This tourist destination is in the Far North Province of what country?
Cameroon

198. The Bandiagara Escarpment is an UNESCO World Heritage Site in the Mopti Region of what country?
Mali

199. Chefchaouen, known for its blue-covered buildings, is in the Rif Mountains, near the Kef Toghobeit Cave, one of the deepest in Africa. This city is near Tangier in what country?
Morocco

200. The Blyde River Canyon, home to Blyderivierpoort Dam, is made up of red sandstone and contains the Three Rondavels, three round mountaintops with pointed peaks in what country?
South Africa

201. The Tundavala Gap, a huge gorge on the Serra da Leba Escarpment, provides spectacular scenic views of the plateau in front of it. This gap is near Lubango, in the Huila Province of what country?
Angola

202. Lake Retba, whose waters are naturally pink, lies north of the Cap Vert Peninsula in what country?

Senegal

203. The unique houses of Tiebele, known for their Gurunsi architecture and intricately decorated walls, are built by the Kassena people in the Nahouri Province of what country?
Burkina Faso

204. The famous town of Franschhoek is home to the Huguenot Memorial Musem, the Franschhoek Valley, and Mont Rochelle Nature Reserve. Known for its vineyards and Cape Dutch architecture, this settlement is one of the oldest towns in what country?
South Africa

205. Dougga, a World Heritage Site, was an ancient Roman city and is home to the Libyco-Punic Mausoleum in the northern part of what country?
Tunisia

206. The Nabiyotum Crater, a geologic wonder, is located in Lake Turkana, the world's largest alkaline lake. This crater is in what country home to Samburu National Reserve, on the Ewaso N'giro River?
Kenya

207. Tsingy de Bemaraha National Park, home to gouged limestone caverns, fissures, and karst formations, is located in the Melaky Region of what island country?
Madagascar

208. Pendjari National Park, named after the Pendjari, is located in northern Benin. This national park borders Arli National Park in what country?
Burkina Faso

209. The Grand Mosque of Bobo-Dioulasso is considered the world's largest example of Sudano-Sahalian architecture. Bobo-Dioulasso, in the Houet Province, is the second largest city in what country?
Burkina Faso

210. Dzanga-Sangha Special Reserve, which is part of a reserve complex with Lobeke National Park in Cameroon and Nouabale-Ndoki National Park in the Republic of the Congo is located in what country?
Central African Republic

211. Odzala National Park is located in the Cuvette-Ouest Region of what Central African country?
Republic of the Congo

212. Name the most populous French-speaking city in West Africa that is the economic capital of Cote d'Ivoire, one of the world's major cocoa producers.
Abidjan

213. Monte Alen National Park, home to a high population of goliath frogs, is the largest national park in what country whose capital is situated on the island of Bioko?

Equatorial Guinea

214. Cape Zebib, near Metline, is in the Bizerte Governorate of what country home to Kairouan and the Mosque of Uqba, considered one of its most important mosques?
Tunisia

215. Tamale, home to Tamale Stadium, is the fourth largest city in what country home to Manhyia Palace in Kumasi, the capital of the former Ashanti Kingdom?
Ghana

216. The Itombwe Mountains are home to a population of the Eastern Lowland Gorilla in what coutry?
Democratic Republic of the Congo

217. Oudtshoorn is known as the "ostrich capital of the world" because it is famous for being the home of the most ostriches in the world. This city is the largest in the Little Karoo region of what country?
South Africa

218. Malindi, home to the Broglio Space Center, is located at the mouth of the Athi-Galana-Sabaki River System and is a major city in what country home to the Loita Plains?
Kenya

219. Copper has been extracted from the 1940s from Musonoi Mine. This mine, with Lake Nzilo, is located near Kolwezi, the capital of Lualaba Province in what country?

Democratic Republic of the Congo

220. The city of Khayelitsha, located between Table Bay and False Bay, is situated on the Cape Flats in what country?
South Africa

221. The famous Mbozi Meteorite and Ruaha National Park are located near Mbeya, in the Mbeya Region and in the southwestern part of what country?
Tanzania

222. Meknes, in the Fes-Meknes Region, was one of the four imperial cities of what country?
Morocco

223. The city of Walvis Bay, home to Kuisebmund Stadium, is situated on the Kuiseb River Delta in what country?
Namibia

224. The confluence of the Calabar River and Cross River is located south of the city of Calabar. The Calabar River drains part of the Oban Hills in what country?
Nigeria

225. Goma, in the Democratic Republic of the Congo is on the northern shore of Lake Kivu, adjacent to the city of Gisenyi. Gisenyi, known for its water sports, is on a flat plain near Mount Nyiragongo in what country?
Rwanda

226. The port of Tanga is the oldest operating port and the most northerly seaport of what country home to Olduvai Gorge, considered one of the most important paleoanthropological sites in the world?
Tanzania

227. The Ruins of Gedi are located within Arabuko Sokoke Forest and National Park, ten miles south of Melindi, on the Galana River in what country?
Kenya

228. Bubaque Island, known for its rich wildlife and forests, is the most populous island in what archipelago of Sierra Leone?
Bissagos Islands

229. Jos, home to Jos Wildlife Park, is known as "J-Town" and is the capital of the Plateau State, on the Jos Plateau. This city is in the Middle Belt region of what country?
Nigeria

230. The artificial island of Eden is a popular tourist destination in the Indian Ocean in the Mahe Islands, which belong to what archipelagic country?
Seychelles

231. Bahir Dar, the capital of the Amhara Region, is located on Lake Tana's southern shore in what country home to Mount Tullu Dimtu, on the Sanetti Plateau?
Ethiopia

232. The city of Bamenda is near Mount Oku and Lake Oku, and is on the Oku Massif. This city, volcano, and lake are located on the Volcanic Line of what country?
Cameroon

233. Lobito, known as the "Flamingo City", is in Benguela Province north of the Catumbela Estuary in what country?
Angola

234. Garamba National Park, one of Africa's oldest national parks, was home to the last surviving population of northern white rhinos in the wild and is in what country?
Democratic Republic of the Congo

235. Fort Jesus is a Portuguese fort situated on what island in Kenya linked to the mainland by the Makupa Causeway, Nyali Bridge, and Likoni Ferry?
Mombasa Island

236. The Chongoni Rock Art Area is located near the town of Dedza, the tallest town in the Central Region and the tallest town in what country?
Malawi

237. The Hanish Islands lie northeast of the Bay of Beylul, a body of water bordering what country?
Eritrea

238. Lake Kompienga is a reservoir that was created by Kompienga Dam in the 1980s and is located in the Kompienga Province of what country?
Burkina Faso

239. Gondar is located north of Lake Tana, and situated on the Lesser Angereb River in what country?
Ethiopia

240. The Moses Mabhida Stadium and the Inkosi Albert Luthuli International Convention Center are located in the city of Durban in what country?
South Africa

Australia and Oceania

NOTE: This chapter can be used to prepare for the **Australian Geography Competition and the Geography's Big Week Out event in Australia.** However, it can also be useful in the National Geographic Bee for any Australia questions.

1. The Brigalow Belt, a wide band of acacia wooded grassland runs across part of what Australian state whose capital is Brisbane?
   Queensland

2. Mount Tomanivi, the highest point in Fiji, is located on what Fijian Island that is the largest in the country?
   Viti Levu

3. Bokak Atoll, located in the Ratak Chain, contains the northernmost point in what country?
   Marshall Islands

4. The Falefa Valley is located on the eastern side of the island of Upolu in what country?
   Samoa

5. Kosrae Island, the easternmost of the Caroline Islands, is located in what country?
   Federated States of Micronesia

6. Raoul Island, the largest and northernmost of the Kermadec Islands, belongs to what Polynesian country?
    New Zealand

7. Lake Gairdner, considered to be the third largest salt lake in Australia, is to the north of the Eyre Peninsula in what Australian state?
    South Australia

8. Jellyfish Lake is a marine lake located on Eli Malk Island, which is part of the Rock Islands archipelago in what country?
    Palau

9. The Finisterre Mountain Range is located on the Huon Peninsula in what Melanesian country?
    Papua New Guinea

10. Espiritu Santo, which belongs to the New Hebrides archipelago, is part of what country's Sanma Province?
    Vanuatu

11. Fakaofo Atoll is located in what territory of New Zealand whose Swains Island is actually administered by the United States in American Samoa?
    Tokelau

12. The Chinese Garden of Friendship, which represents the Ming Dynasty, is located in Chinatown in what major city in Australia?
Sydney

13. Purnululu National Park is a World Heritage Site located near the town of Halls Creek in what Australian state?
Western Australia

14. The Loyalty Islands, which form the Loyalty Islands Province, are northeast of the main island of what French territory?
New Caledonia

15. Bunbury is a city on what major bay in Western Australia?
Geographe Bay

16. Cyclone Bebe hit the island of Funafuti in 1972. This island is located in what Polynesian country?
Tuvalu

17. Buada Lagoon and Anibare Bay are located in what country whose highest point is Command Ridge?
Nauru

18. The Tonda Wildlife Management Area is a wetland in what country home to Kuk Swamp, a UNESCO World Heritage Site?
Papua New Guinea

19. Kavachi Volcano is a submarine volcano located south of Vangunu Island in what country?
Solomon Islands

20. Hawke Bay on North Island borders what peninsula to the east?
Mahia Peninsula

21. What river originates in the Victor Emanuel Range and is the longest river in Papua New Guinea?
Sepik River

22. The You Yangs are granite ridges on the Werribee Plain, partially protected by You Yangs Regional Park. These landforms are northeast of Geelong in what Australian state?
Victoria

23. Cape Kidnappers in New Zealand juts out into the ocean on the eastern coast of what island?
North Island

24. Port Headland is a major city in the northern part of what state in Australia that is the largest by area?
Western Australia

25. Cape Naturaliste juts out into Geographe Bay, on the western coast of what country?
Australia

26. What country's highest point is Mont Panie on the island of Grande Terre?
New Caledonia

27. Alofi Bay and Avatele Bay borders what territory of New Zealand whose westernmost point is Halagigie Point?
Niue

28. The Garden of the Sleeping Giant is an orchid collection on a plantation in what country that occupies the Yasawa Group of islands?
Fiji

29. The Amedee Lighthouse, located near the capital city of Noumea in New Caledonia, is located on what island?
Amedee Island

30. Hamilton Island, the commercial island of the Whitsunday Islands, is located in what Australian state?
Queensland

31. Waipoua Forest preserves an area of breathtaking kauri forest on what island in New Zealand?
North Island

32. The Pinnacles, a set of breathtaking limestone formations in the desert, are located within Nambung National Park in what Australian state?
Western Australia

33. Pegasus Bay, located north of the Banks Peninsula, is east of what island in New Zealand?
South Island

34. The islands of Santa Isabel and Choiseul in the Solomon Islands border what sound to the south?
New Georgia

35. Goyder Lagoon is an ephemeral swamp on the floodplain of the Diamantina River in what country?
Australia

36. Levuka, on Ovalau Island in the Lomaiviti Province, is a UNESCO World Heritage Site in what country home to the Sri Siva Subramaniya Temple in Nadi?
Fiji

37. Bora-Bora, part of the Society Islands, is northwest of Raiatea and Huahine Islands in what French territory?
French Polynesia

38. The Hervey Islands are located southeast of Aitutaki Atoll in what island group that belongs to New Zealand?
Cook Islands

39. The Ngebedech Terraces are in the southwestern part of what country whose capital is Melekeok?
Palau

40. The Sigatoka Sand Dunes form a national park at the mouth of the Sigatoka River in what country?
Fiji

41. Name the only country in the continent of Australia that borders Indonesia.
Papua New Guinea

42. The Papaseea Sliding Rocks can be found on the island of Upolu in what country?
Samoa

43. French and Tahitian are the languages of what island whose capital is Vaitape and is a major international tourist destination, with many luxury resorts?
Bora Bora

44. The islands of Uvea and Iles Wallis belong to what country?
France

45. Dalhousie Springs is located in Witjira National Park in what country?
Australia

46. The Stone Monuments of Mu'a were burial sites for the kings of what country?
Tonga

47. The Torricelli Mountains border the Bewani Mountains to the west and the Prince Alexander Mountains to the east in what country?
Papua New Guinea

48. The Rock Islands, which form the islands in the Southern Lagoon, are part of the Koror State of what Micronesian country?
Palau

49. What country, whose capital's islets are connected by causeways, was formerly known as the Gilbert Islands?
Kiribati

50. Anaa is located in the northwestern part of the Tuamotu Archipelago in what French territory?
French Polynesia

51. Te Urewera National Park, which is home to Lake Waikaremoana, consists of the largest remaining area of native forest on what island in New Zealand?
North Island

52. The Marquesas Islands, which consist of the islands of Tahuata and Nuku Hiva, belong to what Western European country?
France

53. The Phoenix Islands and the Line Islands belong to what country?

Kiribati

54. The Cobourg Peninsula, located east of Melville Island, is in what territory of Australia?
Northern Territory

55. What island territory, sometimes referred to as the Union Islands, is home to the Luana Liki, the only hotel in the territory?
Tokelau

56. Lake Hiller, a pink saline lake, is located on Middle Island, the largest island in what Australian archipelago home to Cape Arid National Park?
Recherche Archipelago

57. Boomerang Beach is located miles south of the city of Taree in what southeastern Australians state?
New South Wales

58. Whanganui National Park, which borders the Whanganui River, is in what country bordering the Tasman Sea?
New Zealand

59. Adelaide is the capital of what Australian state?
South Australia

60. You can buy and eat pavlovas at cafes in what country home to the Coromandel Peninsula and the Colville Channel?

New Zealand

61. Kangaroo Island is located west of Lacepede Bay in what Australian state?
South Australia

62. The Caves of Nanumanga are underwater caves off the northern coast of Nanumanga, a reef island and district of what Polynesian country?
Tuvalu

63. Aore Island is opposite to what city on Espiritu Santo Island in Vanuatu?
Luganville

64. Marshall Islands International Airport is on what island home to the largest city in the country?
Majuro Island

65. The Tanami Desert stretches across the western region of what territory in Australia bordering Joseph Bonaparte Gulf?
Northern Territory

66. The Port Perpendicular Lighthouse is on the Beecroft Peninsula. This peninsula juts out into Jervis Bay in what Australian state?
New South Wales

67. Rota and Alamagan are islands in what Micronesian territory of the United States?
Northern Mariana Islands

68. Tamborine Mountain and the Numinbah Valley are part of the Gold Coast hinterland in what Australian state?
Queensland

69. Pohnpei, home to the city of Palikir, is located in what island group?
Senyavin Islands

70. Cape Farewell extends across Golden Bay on what island in New Zealand?
South Island

71. Lake Ellesmere, which borders Canterbury Bight, is located west of what peninsula on South Island?
Banks Peninsula

72. The Vaipo Waterfall is on the island of Niku Hiva, the largest island in what archipelago in French Polynesia?
Marquesas Islands

73. Lamington National Park, in the McPherson Mountain Range, is on the border between Queensland and what other Australian state?
New South Wales

74. Nitmiluk National Park, located around a series of gorges on the Katherine River and Edith Falls, borders Kakadu National Park in what country?
Australia

75. The Rabaul Caldera, a volcano on the tip of the Gazelle Peninsula, is located in what country?
Papua New Guinea

76. A Shri Shiva Vishnu Temple can be found in what Australian state whose capital is Melbourne?
Victoria

77. Tavurvur, an active stratovolcano, is located on what island in Papua New Guinea that is the largest in the Bismarck Archipelago?
New Britain Island

78. The Great Papuan Plateau, located in the South Highlands Province in Papua New Guinea, can also be found in what other province?
Western Province

79. Hufangalupe is a natural land bridge on what island in Tonga home to the country's capital, Nuku'alofa?
Tongatapu

80. The Daintree Rainforest is a tropical region north of the Daintree River and in Daintree National Park. It is also north of Mossman and Cairns in what Australian state?

Queensland

81. The Miramar and Kaikoura Peninsulas are in what country home to the Karangahake Gorge, between the Coromandel and Kaimai Ranges?
New Zealand

82. Ball's Pyramid, part of Lord Howe Island Marine Park, is the tallest volcanic stack in the world. This remnant of a shield volcano and caldera is located twelve miles southeast of Lord Howe Island in what country?
Australia

83. The Hamersley Range, in the Pilbara, is home to Karijini National Park, one of Australia's largest national parks. This range contains Mount Meharry, the highest point in what Australian state?
Western Australia

84. What lake in New Zealand, drained by the Waikato River, is in the Waikato Region on North Island and lies in the volcano caldera of the same name?
Lake Taupo

85. Momote Airport, located on Los Negros Island, is in what island group in Papua New Guinea?
Admiralty Islands

86. The Mamanuca Islands are a volcanic island group south of the Yasawa Islands and west of Nadi in what country?

Fiji

87. Koonalda Cave, on the Nullarbor Plain, is situated within the Nullarbor Wilderness Protection Area. This cave is near Ceduna, on Murat Bay in what Australian state?
South Australia

88. The Tookoonooka Crater is located in the Eromanga Basin, which extends into part of the Cooper Basin. This crater is in what Australian state?
Queensland

89. The Illawarra Escarpment, whose highest point is Bells Hill, is south of what major city surrounding one of the world's largest natural harbors?
Sydney

90. The Shortland Islands and the Russell Islands belong to what country east of Papua New Guinea?
Solomon Islands

91. Lake Kutubu, east of the Kikori River, is southwest of Mendi, the capital of the Southern Highlands Province. This lake is the second largest in what country?
Papua New Guinea

92. Vai Lahi Crater Lake is in the central part of Niuafo'ou Island, near Niuatoputapu Airport in what country home to Hunga Tonga Volcano in the Ha'apai Archipelago?
Tonga

93. The Falemauga Caves are located on Upolu Island in what country whose capital is Apia?
Samoa

94. Opunohu Bay is located on the island of Moorea near Tahiti in what archipelago?
Society Islands

95. Vanikoro, which is part of the Temotu Province, is located in what island group in the Solomon Islands?
Santa Cruz Islands

96. The Dorrigo Plateau, in the Northern Tablelands of New South Wales, forms part of what mountain range home to the Barren Mountain?
Great Dividing Range

97. Piccaninnie Ponds Conservation Park is home to wetlands and freshwater springs near Mount Gambier and Discovery Bay in what Australian state?
South Australia

98. Mount Yasur, an active volcano located near Sulphur Bay, is located on Tanna Island in what country?
Vanuatu

99. The Three Sisters are a unique rock formation in the Jamison Valley of what mountain range of New South

Wales home to Wollemi National Park and Wentworth Falls?
Blue Mountains

100. Rangiroa is the largest atoll in what island group in French Polynesia?
Tuamotu Archipelago

101. Lake Ngardok, the source of the Ngerdorch River, is located in the State of Melekeok in Palau on what island?
Babeldoab

102. The Top End is a geographical region comprising the northernmost part of what Australian administrative division?
Northern Territory

103. The Koro Sea, which borders the Lau Islands to the east, is surrounded by what country?
Fiji

104. The Moqua Well is a small underground lake in what country bordering Anibare Bay?
Nauru

105. There is a meteorological observatory on what islet in the atoll of Funafuti in Tuvalu?
Fongafale Observatory

106. Amatuku Islet, which is located at the end of the Tengako Peninsula, is located in what country?
Tuvalu

107. Ked Ra Ngchemiangel, also known as the Kamyangel Terraces, are landforms near the city and state of Aimeliik, on the Babeldoab in what country?
Palau

108. Lake Corangamite, which is located near Colac, is in what country whose Becher Point Wetlands are located along the Swan Coastal Plain?
Australia

109. Wilpena Pound, a geographical wonder with numerous rock structures, gorges, and mountains is in Flinders Range National Park in what Australian state?
South Australia

110. Undara National Park is in the Einasleigh Uplands in what Australian state home to the Cape York Peninsula?
Queensland

111. Ongael Lake, known by its Palauan name of "Uet era Ongael", is a marine lake in what Palauan state once the capital of the South Pacific Mandate of Japan following World War I?
Koror

112. The Yorke Peninsula, located between Spencer Gulf and Gulf St. Vincent, is separated from Kangaroo Island by what strait?
Investigator Strait

113. The Klee Passage separates Knox Atoll from Mili Atoll in what country?
Marshall Islands

114. Mont Orohena, the highest peak in French Polynesia, is located on what island?
Tahiti

115. Chuuk Lagoon, also known as Truk Lagoon, is part of the Chuuk State in what country made up of four states?
Federated States of Micronesia

116. The Whangaparaoa Peninsula is part of the Hibiscus Coast on what island in New Zealand?
North Island

117. Kings Canyon is part of Watarrka National Park in what administrative division of Australia?
Northern Territory

118. Wollomombi Falls is located in Oxley Wild Rivers National Park and is a plunge waterfall in what region of New South Wales?
New England

119. Murphy's Haystacks are inselberg or monadnock rock formations situated on the Eyre Peninsula in what Australian state?
South Australia

120. The Vava'u Islands, whose capital is Neiafu, are home to the 'Ene'io Botanical Garden and are located in what Polynesian archipelagic country?
Tonga

121. Tarkine is a famous wilderness area home to Savage River National Park on what Australian island?
Tasmania

122. The Louisiade Archipelago is a group of volcanic and coral islands located southeast of what country?
Papua New Guinea

123. You can view the islands of the Recherche Archipelago from Lucky Bay, which borders Cape Le Grand National Park in what Australian state?
Western Australia

124. The Ahukawakawa Swamp is located in Egmont National Park, in the Taranaki Region of what country?
New Zealand

125. Lake Callabonna, a dry salt lake in the northern part of South Australia, is located 75 miles southwest of what

point that is the junction of the state borders of Queensland, South Australia, and New South Wales?
Cameron Corner

126. The Bogong High Plains, located in the Victorian Alps, can be found in the Australian state of Victoria in what national park?
Alpine National Park

127. The Cronulla Sand Dunes are located on the Kurnell Peninsula in what southeastern Australian state?
New South Wales

128. The largest island in the Trobriand Islands is Kiriwina Island, which is part of what country?
Papua New Guinea

129. The Louisiade Archipelago is located in the Milne Bay Province of what Melanesian country?
Papua New Guinea

130. Resolution Island, which is home to a peninsula on its west coast called the Five Fingers Peninsula, borders Dusky Sound and Breaksea Sound in what country?
New Zealand

131. The North Island Volcanic Plateau and Pureora Forest Park are located in what country?
New Zealand

132. Joske's Thumb is a volcanic plug on what island in Fiji that is home to the country's capital city, Suva?
Viti Levu

133. Frying Pan Lake, one of the world's largest hot springs, was formed from the enormous Mount Tarawera eruption in 1886, and is located near Inferno Crater Lake in the Waimangu Volcanic Rift Valley in what country?
New Zealand

134. The Milne Bay Province belongs to what large Melanesian country?
Papua New Guinea

135. Fiordland National Park is located in the southern part of what country?
New Zealand

136. The majority of Australian Hindus live in Sydney and what other city home to a Hindu Murugan Temple and a war memorial known as the Shrine of Remembrance?
Melbourne

Antarctica

1. Name the longest river in Antarctica, whose source is at Lake Brownworth in Victoria Land.
    Onyx River

2. Jang Bogo Station, located on Terra Nova Bay, belongs to what country on the Korean Peninsula?
    South Korea

3. Mount Hampton, which is part of the Executive Committee Mountain Range, is a volcano in what region of Antarctica?
    Mary Byrd Land

4. Deacon Peak is the highest point on what island in the South Shetland Islands?
    Penguin Island

5. Spaatz Island lies west of the southwestern part of the Antarctic Peninsula and southwest of what island, the largest in Antarctica?
    Alexander Island

6. Thurston Island lies on the western half of what ice shelf in Antarctica?
    Abbot Ice Shelf

7. The Bharati Research Station, located in the eastern part of Antarctica, belongs to what country in South Asia?
    India

8. Maldonado Base, situated on Guayaquil Bay at Greenwich Island in the South Shetland Islands, was named after an astronomer and geographer from Riobamba in what country in South America?
    Ecuador

9. What Indian research station in Antarctica is situated on the Schirmacher Oasis, an ice-free plateau consisting of freshwater lakes on the Princess Astrid Coast in Queen Maud Land?
    Maitri Station

10. Mount Melbourne, an enormous stratovolcano, is located between Wood Bay and Terra Nova Bay in what region of Antarctica?
    Victoria Land

11. Paulet Island, which lies southeast of Dundee Island, is located off the northeastern coast of what peninsula?
    Antarctic Peninsula

12. The Larsemann Hills extend from Dalk Glacier and are situated on the southeastern coast of what bay bordering the Amery Ice Shelf?
    Prydz Bay

13. The McMurdo Dry Valleys, which consist of snow-free valleys, is home to the Onyx River and what hypersaline lake in the Victoria Valley that is the northernmost in the dry valley region?
Lake Vida

14. The Pagodroma Gorge, which connects the Radok and Beaver Lakes, is in what mountain range whose highest peak is Mount Menzies?
Prince Charles Mountains

15. The Olympus Range borders the Wright Valley, which is west of what sound covered by an ice shelf of the same name?
McMurdo Range

16. The Haskell Strait is located between Cape Spencer-Smith on White Island and Cape Armitage on what peninsula on Ross Island?
Hut Point Peninsula

17. The Dyer Plateau borders the Gutenko Mountains to the south in what region in the Antarctic Peninsula?
Palmer Land

18. The Ritscher Upland is situated to the east of the Kraul Mountains near what glacier?
Veststraumen Range

19. The Trueman Terraces are near what mountain range whose highest point is Holmes Summit and is located between the Recovery and Slessor Glaciers?
    Shackleton Range

20. The Nishino-seto Strait is located between Ongulkalven Island and Ongul Island in the Flatvaer Islands in what bay bordering the Riiser-Larsen Peninsula?
    Lutzow-Holm Bay

21. Alexander Island is separated from the Antarctic Peninsula by George VI Sound and what other body of water?
    Marguerite Bay

22. Name the southernmost active volcano on Earth, located on Ross Island in Antarctica.
    Mount Erebus

23. Prime Head is the northernmost point on the Antarctic mainland, and is situated on what peninsula?
    Trinity Peninsula

24. What bay is between Point Alden and Cape Gray and whose famous storms are caused by katabatic winds?
    Commonwealth Bay

25. Lake Untersee is located nearly sixty miles southwest of the Schirmacher Oasis and is a large surface freshwater lake in what mountain range in Queen Maud Land that is part of the Wohlthat Mountains?

Gruber Mountains

26. Cape Royds is located on Ross Island and juts out into what inlet?
McMurdo Range

27. The Linnaeus Terrace is situated on the northern side of Oliver Peak in what mountain range?
Asgard Range

28. The Langhovde Hills are home to the Yukidori Valley on what bay's eastern coast?
Lutzow-Holm Bay

29. Dakshin Gangotri Glacier was named for a scientific base station built by what country that commissioned the Bharati Research Station?
India

30. The Marion Nunatacks are situated on Charcot Island, near Cape Byrd in what sea?
Bellingshausen Sea

India

NOTE: This chapter is only to prepare for the **Regionals and Nationals of the North South Foundation (NSF) Junior and Senior Geography Bees.** However, it can also be useful in the National Geographic Bee for any India questions.

1. Agasthyarkoodam is a peak in the Neyyar Wildlife Sanctuary. This landfom, part of the Agasthyamala Biosphere Reserve, is in what Indian state?
    Tamil Nadu

2. What city, known as the "Detroit of South Asia" for its successful automobile industry, is home to Valluvar Kottam?
    Chennai

3. What city in South India is known as the "Silicon Valley of India", and is home to the headquarters of Infosys?
    Bengaluru

4. Mudumalai National Park, home to endangered Bengal tigers, Indian elephants, and Indian leopards, lies on the northwestern side of the Nilgiri Hills and shares its boundaries with the states of Karnataka and Kerala. This national park is in what state?

Tamil Nadu

5. Shillong, known as the "Scotland of the East", is the capital of what state?
   Meghalaya

6. Indravati National Park, which derives its name from the Indravati River, is located in the Bijapur District of what state home to Chitrakote Falls?
   Chhattisgarh

7. Connaught Place is a major financial, commercial, and business center in what metropolitan area on the Yamuna River?
   Delhi

8. The Meenakshi Amman Temple, on the southern banks of the Vaigai River, is in what city, the third largest in Tamil Nadu?
   Madurai

9. Vidyasagar Setu is a bridge connecting Kolkata to Howrah that was built over what river?
   Hooghly River

10. The Bandra-Worli Sea Link connects the suburb of Bandra to the locality of Worli in what major city in Maharashtra?
    Mumbai

11. The Lingaraj Temple, which represents Kalinga architecture, is one of the oldest temples in what city lying southwest of the Mahanadi River in Eastern India?
Bhubaneswar

12. The Pushkar Fair is an annual five-day camel and livestock fair held in the town of Pushkar in what state?
Rajasthan

13. The Papi Hills are a scenic gorge located on the Godavari River in what state?
Andhra Pradesh

14. Puri, Konark, and what other Orissan city form the Swarna Tribhuja, or the "Golden Triangle", one of eastern India's most visited destinations?
Bhubaneswar

15. The Vembanad Rail Bridge connects Edappally and Vallarpadam in what major city in Kerala?
Kochi

16. Sri Venkateswara National Park is a national park and biosphere reserve, home to the Talakona Waterfalls, in the Chittoor District of what South Indian state bordering Telangana?
Andhra Pradesh

17. The Harmandir Sahib, known as the "Golden Temple", is the holiest Gurudwara of Sikhism, and is in what city?

Amritsar

18. The Jagannath Temple of Puri is on the eastern coast of India and is famous for its Rath Yatra, or chariot festival. This temple is in what state bordering the Bay of Bengal?
Odisha (Orissa)

19. The Kapaleeshwarar Temple and the Mahishasuramardini Mandapa are sites in what South Indian state?
Tamil Nadu

20. Biscomaun Bhawan, the tallest building in Bihar, is located in what city?
Patna

21. The Ekambareswarar Temple is the largest temple in Kanchipuram, on the banks of the Vegavathy River in what Indian state?
Tamil Nadu

22. Pattadakal, located on the left bank of the Malaprabha River, is a World Heritage Site in what Indian state?
Karnataka

23. The Chhatrapati Shivaji Terminus railway station is in what major city on Salsette Island?
Mumbai

24. What city, at the southern edge of the Malwa Plateau, is the largest city in Madhya Pradesh?
Indore

25. Jantar Mantar is home to the world's largest sundial, a device telling the time of day by the sun's apparent position in the sky. This World Heritage Site, built in 1734, can be found near the Hawa Mahal in what city?
Jaipur

26. Khangchendzonga National Park is home to the Zemu Glacier and populations of musk deer, snow leopard, and Himalayan tahr, in what Indian state?
Sikkim

27. The Ganesha Temple at Morgaon, also known as the Shri Mayureshwarar Mandir, is fifty miles away from Pune in what Indian state?
Maharashtra

28. The Open Hand Monument is a symbolic structure in what Indian union territory?
Chandigarh

29. The New Yamuna Bridge, built over the Yamuna River, is in what city that hosted the Kumbh Mela in 2013?
Allahabad

30. Valley of Flowers National Park is home to rare and endangered species such as the Asiatic black bear and

the snow leopard. This World Heritage Site is in the Western Himalaya of what state?
Uttarakhand

31. The Penumudi-Puligadda Bridge, located in Andhra Pradesh, was built over what river?
Krishna River

32. The Shankaracharya Temple, situated on top of Shankaracharya Hill, is in what city that is the summer capital of Jammu and Kashmir?
Srinagar

33. InfoPark is an information technology park in what city that is the most populous metropolitan area in Kerala?
Kochi

34. Ran ki vav, an intricately constructed stepwell in the Patan, is a World Heritage Site in what Indian state?
Gujarat

35. Kanyakumari, the southernmost point in mainland India, is at the southern tip of what subrange and extension of the Western Ghats in Tamil Nadu?
Cardamom Hills

36. The Victoria Memorial is located in what major city on the banks of the Hooghly River in West Bengal?
Kolkata

37. The Brihadeeswarar Temple is one of the largest temples in India, built during the Chola period. This World Heritage Site is part of the "Great Living Chola Temples" and is in what city?
Thanjavur

38. Nehru Setu, the second longest railway bridge in India, is located across the Son River in what state?
Bihar

39. The Laxminarayanan Temple attracts many people during the festival of Diwali in what major city with the Akshardham Temple?
Delhi

40. Gir Forest National Park is home to a little over 520 Asiatic Lions, the last of their species and the only population left on Earth. This national park is a forest and wildlife sanctuary in what Indian state?
Gujarat

41. The Statue of Mahatma Gandhi, in Gandhi Maidan, is near the Ganges River and is the world's tallest bronze statue of Mahatma Gandhi. This site is in what city?
Patna

42. What city in Tamil Nadu lies on the banks of the Noyyal River and is home to the Perur Pateeswarar Temple?
Coimbatore

43. What city in Haryana is known as the "City of Steel" and is also home to Blue Bird Lake?
Hisar

44. Jaisalmer Fort, one of the largest fortifications in the world, is a UNESCO World Heritage Site in Jaisalmer in what Indian state?
Rajasthan

45. Lodi Gardens, home to the Tomb of Sikandar Lodi, is a city park in what Indian city?
New Delhi

46. Guwahati, the capital of Assam, is situated on what major Indian river?
Brahmaputra

47. Thekaddy, the site of Periyar National Park, is a tourist attraction in what Indian state?
Kerala

48. The Rohtang Pass is a high mountain pass in Himachal Pradesh, in what subrange of the Himalayas?
Pir Panjal Range

49. Ramoji Film City, the largest integrated film city in the world, is located in what major Indian city?
Hyderabad

50. Abbey Falls, located in the Western Ghats, is a site in Kodagu in what Indian state?

Karnataka

51. The Solang Valley, a side valley at the top of the Kullu Valley, is located in what mountainous Indian state?
Himachal Pradesh

52. The Great Stupa at Sanchi is the oldest stone structure in India. This Buddhist vihara and UNESCO World Heritage Site is in what Indian state?
Madhya Pradesh

53. Sipahijola Wildlife Sanctuary, near Bishalgarh, is in what Indian state?
Tripura

54. The Rohtas Plateau is in the southwestern part of what Indian state whose capital lies at the confluence of four rivers?
Bihar

55. The Marble Rocks, located along the Narmada River, are a gorge of white marble and are rich in magnesium. They can be found near Jabalpur in what Indian state?
Madhya Pradesh

56. The Kallil Temple is a Jain temple near Kalady in what Indian state?
Kerala

57. The Langpangkong Mountain Range, inhabited by the Ao people, is located in what Indian state home to the Mokokchung District?
Nagaland

58. The Bhitarkanika Mangroves, a mangrove wetland in the river delta of the Brahmani and Baitarani Rivers, are in what Indian state?
Odisha (Orissa)

59. Kollam is located on Ashtamudi Lake, a lake in the backwaters of what Indian state?
Kerala

60. Muthathi, on the banks of the Kaveri River, is surrounded by Kaveri Wildlife Sanctuary in what Indian state?
Karnataka

61. Kutladampatti Falls is located northwest of Madurai in what Indian state?
Tamil Nadu

62. The Harike Wetland and Lake are located near the confluence of the Sutlej and Beas Rivers in what Indian state bordering Rajasthan?
Punjab

63. The Pallikaranai Wetland is located south of the Adyar River and is one of the few and last remaining wetlands in Southern India. This wetland region, along with the

Vedanthangal Bird Sanctuary, is home to many species of birds and is in what Indian state?
Tamil Nadu

64. The Barabar Caves, the oldest surviving rock-cut caves in India, date from the Mauryan Empire and some have been inscribed with Ashokan writing. These caves are north of Gaya in what Indian state?
Bihar

65. Kanger Ghati National Park is located near Jagdalpur. Jagdalpur, home to the Kotumsar Cave, is on the Indravati River in what Indian state?
Chhattisgarh

Physical Geography/World of Science

NOTE: This round can also be used to prepare for the **Geomorphology** and **Climate Geography** questions of the **United States Geography Olympiad (USGO).** This chapter can also help **Ocean Wonders** rounds in the National Geographic Bee (a major category in 2017).

1. What term describes a type of broad and rounded cliff bordering a coastal area like the feature found in Minnesota, carved out by the Mississippi River?
    Bluff

2. What is the term that describes a large depression resulting from the collapse of the center of a volcano?
    Caldera

3. What is the term for a narrow strip of land connecting two larger landmasses?
    Isthmus

4. A point of land smaller than a peninsula that juts out into a body of water is known by what term?
Cape

5. A tableland with a flat top and steep cliffs around the edge found in arid regions and larger than a butte is known by what name?
Mesa

6. What term describes a depression in the ground caused by the Earth's crust spreading apart or tectonic plates spreading apart?
Rift Valley

7. A flat, low-lying plain that sometimes forms at the mouth of a river from deposits of sediments.
Delta

8. A body of water partially surrounded by land, usually with a wide mouth to a larger body of water.
Bay

9. A stream that feeds or flows into another stream or river.
Tributary

10. A large area of water lying within a curved coastline, usually larger than a bay.
Gulf

11. What conical volcano is also known as a composite volcano and are one of the most common type of volcanoes in the world?
Stratovolcano

12. A raised fault block surrounded by grabens or normal faults is known by what term?
Horst

13. What Irish term describes a disappearing lake found in limestone areas of Ireland?
Turlough

14. What term describes the mouth of a river where the river's current meets the sea's tide?
Estuary

15. What is the term that was derived from Kazakh and describes a crescent-shape dune, especially in inland desert regions?
Barchan (Barkhan)

16. A flat and narrow area along a shoreline or at the base of a sea cliff created by wave erosion are known by what term?
Wave-cut platform (Coastal bench or Wave-cut bench)

17. An artificial lake, usually created by a dam, is known by what name?
Reservoir

18. What term describes the saltwater surrounding landmasses that is divided by the landmasses into portions?
Ocean

19. What is the term for a place where two rivers or streams join to become one river or stream?
Confluence

20. Bodies of water surrounded by land that are deeper than ponds and are not part of the ocean are known by what name?
Lake

21. An igneous rock or silicate mineral primarily composed of magnesium and iron, such as basalt, is known by what term?
Mafic

22. What term describes a stream that branches off and flows away from a main stream channel?
Distributary

23. A narrow body of water separating two landmasses, similar to a channel or a passage.
Strait

24. A sparsely inhabited region, usually unique to Australia, but without much vegetation, is known by what term?

Bush

25. The large-scale ocean circulation driven by density gradients worldwide by surface heat is known by what name?
Thermohaline Circulation

26. What term describes a long, narrow inlet with steep sides and cliffs, created by glacial erosion?
Fjord

27. A large sea or ocean inlet that is larger than a bay, deeper than a bight, and wider than a fjord is known by what term?
Sound

28. What term describes a large natural flow of what is usually freshwater that crosses an area of land and flows into a body of water, or a natural stream of water?
River

29. A line joining two equal points of rainfall on a map is known by what name?
Isohyet

30. What term describes the boundary between the Earth's crust and mantle?
Mohorovicic Discontinuity (Moho)

31. What is the term describing a common igneous rock formed from the quick cooling of lava at or near the surface of the Earth?
Basalt

32. The occurrence of seasonal changes in atmospheric circulation and precipitation is known by what name?
Monsoon

33. What term, also called a basalt detachment fault, describes the gliding plane between two rock masses?
Decollement

34. What term describes the study of the origin, structure, and composition of rocks?
Petrology

35. Name the type of uncommon shield volcano formed from explosive volcanoes and often has a central caldera.
Pyroclastic Shield (Ignimbrite Shield)

36. What term describes a firm hydraulic structure built from an ocean shore that limits the flow of water and the movement of sediment?
Groyne

37. Igneous rocks abundant in elements that produce quartz and feldspar are known by what term?
Felsic

38. What is the term for a line of bold cliffs?
    Palisades

39. Agricultural land that is plowed or tilled but left unseeded during a growing season is known by what term?
    Fallow

40. What term describes partially decomposed organic soil material?
    Humus

41. What term describes a lake filled by sediment and may leave fertile land or become a wetland?
    Lacustrine Plain

42. What is the term that describes a time of widespread glaciation?
    Ice Age

43. Name the layer of any body of water where the density is the greatest.
    Pycnocline

44. What term refers to a long narrow ridge or mound of sand, gravel, and boulders deposited by a stream flowing on, within, or beneath a stagnant glacier?
    Esker

45. What is the term for a dry terrain where soils consisting of clay and sedimentary rocks have eroded a lot and there is very low vegetation along with steep landforms?
Badlands

46. What is the name of a mostly flat rocky surface divided into regular or irregular rectangles/polygons by fractures or systematic joints in the rock?
Tessellated Pavement

47. A dome-shaped or steep-sided bald-rock outcropping that is at least one hundred feet tall and several hundred meters wide is known by what term?
Bornhardt

48. What is the term for a rock formation created by the passing of a glacier, which often results in asymmetric erosional forms?
Roche moutonnee (Sheepback)

49. What Malagasy term describes a common erosional feature in Madagascar and sometimes in other parts of the world usually found on the side of a hill?
Lavaka

50. What term describes the section of a floodplain where fine silts and clays stay after a flood?
Backswamp

51. What is the term for a table-top mountain or mesa that can be found in the Guiana Highlands of South America and are usually made up of Precambrian quartz arenite sandstone?
Tepui

52. What term describes an elongated or oval-shaped hill formed by glacial drift?
Drumlin

53. Name the landform that is an example of a nunatak and is a sharply-pointed mountain peak created from cirque erosion by diverging glaciers.
Pyramidal Peak

54. What is the name derived from German for the crevasse that forms when moving ice from a glacier separates from the unmoving ice above?
Bergschrund

55. A lake that flows into two separate drainage basins with a drainage divide in the middle of a lake is known by what name?
Bifurcation Lake

56. What is the name of an arch-shaped fold whose core consists of old rocks?
Anticline

57. What is the study of the topographic relief of mountains or other landforms using techniques such as elevation and ordering of a river's tributaries?
Orography

58. A line connecting points of constant or potential density on a map is known by what term?
Isopycnal

59. Active fault scarps can form when an earthquake changes the ground's elevation, so they are usually connected to what type of displacement?
Earth Crustal Displacement (Tectonic Displacement)

60. Isostasy is the gravitational equilibrium between the mantle and what other layer of Earth containing the lithosphere?
Crust

61. In geology, what is another name for the lower layer of the Earth's crust, made up of rocks consisting of magnesium silicate minerals?
Sima Layer

62. What is water called when it has more salinity than freshwater but less than seawater?
Brackish Water (Briny Water)

63. A group of fluvial landforms in a river is known by what name?

River Archipelago

64. The processes that cause the erosion of Earth's surface leading to decreasing elevations are known by what term?
Denudation

65. A type of rock or ventifact that forms in often desert environments because of the action of blowing sand often is in a pyramidal shape. Name this ventifact.
Dreikanter

66. In oceanography, any large system of circulating ocean currents, often involved with a lot of wind movement, is known by what name?
Gyre

67. What Spanish term describes a ridge or a hill with a steep slope on one side and a gentle slope on the another, where harder sedimentary rock is found above a softer layer?
Cuesta

68. A steeply sloping triangular landform created by the differential erosion of a rock is known by what geomorphological name?
Flatiron

69. What is another name for a monadnock, an isolated and elevated landform rising directly up from a relatively flat or gradually sloping plain?
Inselberg

70. What is the geomorphological term describing an irregularly shaped mound or hill made up of a sedimentary rock that builds up in the depression of a retreating glacier and is often associated with kettles?
Kame

71. What type of geological formations are created by mud, slurries, gases, and water, and although volcanoes, produce no lava?
Mud Volcano (Mud Dome)

72. An area of land surrounded by young lava flows is known by what geological term that originated in the Hawaiian language?
Kipuka

73. Rocks containing silicate and aluminum minerals can be found in the upper layer of the Earth's crust. Name this layer, known by a geochemical term.
Sial Layer

74. What term describes a large rock outcrop rising from the gradual slopes of a hill or ridge that is used as the name of some hills in the southwestern United Kingdom?
Tor

75. Name the Slavic word for a large plain of karst that gives its name to a famous karst field in Bosnia and Herzegovina.
Polje

76. Connected cave-like features found in rock such as limestone and sandstone can be found in a large quantity in intertidal areas and deserts are known by what name?
Tafoni

77. What forms the border between the upper and lower continental crusts, is not found in oceanic regions, and is the area at which seismic wave velocity increases discontinuously?
Conrad Discontinuity

78. What term describes mudstone with varying amounts of silt and clay, with the primary mineral being calcite?
Marl

79. What type of wetland accumulates peat and occurs where the water on the surface of the ground is acidic?
Bog

80. Tuya may be derived from an Athabaskan word, and it describes a flat-topped and steep subglacial volcano formed when lava erupts through an ice sheet or glacial

ice mass. This term can describe Herðubreið, a tuya and volcano in the highlands of what country?
Iceland

81. In structural geology, a downward or synformal fold with younger layers to the center of the structure is known by what name?
Syncline

Cultural Geography/Culture Connection

1. What city in South Korea that is home to the Woljeongsa Buddhist Temple will host the 2018 Winter Olympics?
Pyeongchang

2. The Leshan Giant Buddha depicts Maitreya, a future Buddha who will teach pure dharma, according to tradition. This statue, the largest stone Buddha in the world, is at the confluence of the Minjiang, Dadu, and Qingyi Rivers in what Chinese province?
Sichuan Province

3. The San Petronio Basilica, located in Piazza Maggiore, is located in what Italian city home to the Stadio Renato Dall'Ara, a multipurpose stadium?
Bologna

4. Estadio Monumental, also known as Monumental Stadium, is located in what major city in Peru?

Lima

5. Soccer City, a nickname for the First National Bank Stadium, is located in what city that is the capital of Gauteng Province in South Africa?
Johannesburg

6. Pozole is a traditional soup or stew made from hominy and can be seasoned or garnished with chile peppers, onion, and garlic, to mention a few. This celebratory dish is typical in the states of Zacatecas and Morelos in what Latin American country?
Mexico

7. The Mosque of Omar Ibn al-Khattab is the largest mosque in Latin America. This mosque is located in Maicao, which is situated on the Guajira Peninsula in what country?
Colombia

8. Mille-Feuille is made up of layers of puff pastry and pastry cream, but occasionally whipped cream and jam are used as well. This food originated from what country home to the Pays de la Loire Region?
France

9. Wat, also known as Tsebhi in Tigrinya, is a type of curry or stew filled with meat, vegetables, and often spice mixtures, notably berbere. This dish is popular in the

cultures of Ethiopia and what other country on the Horn
of Africa?
Eritrea

10. The Maya Devi Temple is an ancient Buddhist temple and
site considered to be the sacred birthplace of Siddhartha
Gautama. This temple is located in the Buddhist
pilgrimage site of Lumbini in what country?
Nepal

11. Dhammayangyi is a Buddhist temple in Bagan, a city
home to over 2,000 Buddhist sites in what country?
Myanmar

12. The Itsukushima Shrine is a Shinto place of worship in
Hatsukaichi, on Itsukushima Island in what body of water
that connects the Pacific Ocean to the Sea of Japan?
Seto Inland Sea

13. Camp Nou Stadium, the largest stadium in Spain by
capacity, is located in what city in Catalonia home to the
concert hall of Palau de la Musique Catalana?
Barcelona

14. Tetum, one of the official languages of East Timor along
with Portuguese and is part of what language family
from which the language of Rapa Nui, spoken on Easter
Island, evolved?
Austronesian Languages

15. The Mingrelian Language, also known as Iverian, is spoken in the western part of what country in the Caucasus region of Asia?
Georgia

16. The Sheikh Lotfollah Mosque is in Naqsh-e Jahan Square in what Iranian city?
Isfahan

17. The malimbe is a xylophone-like instrument originating from what huge sedimentary basin in equatorial Africa?
Congo Basin

18. The Iraivan Hindu Temple is situated next to the Wailua River, which originates near Mount Waialeale on what Hawaiian Island?
Kauai

19. Bògòlanfini is a type of handmade cotton fabric originating from what West African country that hosted the 2002 African Cup of Nations?
Mali

20. Borg el Arab Stadium is located west of Alexandria and is the largest stadium in what country?
Egypt

21. Igbo is a minority language of what country that is the only sovereign state in Africa to have Spanish as its official language?

Equatorial Guinea

22. The Luba-Kasai language is one of the four official languages of what country with a short coastline on the Atlantic Ocean?
Democratic Republic of the Congo

23. The Bashkir language is spoken in the Republic of Bashkortostan in what country?
Russia

24. The Bavikonda Complex is home to stupas and is a Buddhist monument in what South Indian state?
Andhra Pradesh

25. Westfalenstadion is a football stadium in Dortmund in what German state on the Westphalian Lowland?
North Rhine-Westphalia

26. The Imam Reza Shrine, the world's largest mosque by dimension, is located in what Iranian city that is the capital of the Razavi Khorasan Province and is known as the "City of Paradise"?
Mashhad

27. The Hagia Sophia, once the world's largest church building for a thousand years until the completion of the Seville Cathedral, later became an imperial mosque and is now a museum visited by millions of people every year. This site is located in what country?

Turkey

28. Lingala is a national language of a country where the soukous is a popular dance style. Name this country, northeast of Angola.
Democratic Repulic of the Congo

29. The Ladin language is spoken by people in the Dolomite Mountains and the Belluno Province of what country?
Italy

30. Baleadas are a traditional dish consisting of a flour tortilla filled with fried beans and often cheese. They originate from what country home to the Cathedral of Amapala?
Honduras

31. Sewu, a Mahayana Buddhist temple, is on what Indonesian island where Sundanese and Madurese are spoken along with Javanese?
Java

32. Chewa is a Bantu language with official status Zimbabwe and what other country whose name derives from a kingdom that existed in the 16th and 17th centuries?
Malawi

33. Coconuts are used in Hindu rituals in a country that celebrates Krishna Janmashtami annually along with Nepal. Name this country.

India

34. The Romance language of Extramaduran is spoken by hundreds of thousands of people in parts of the Extramadura Autonomous Community and Salamanca Province of what country?
Spain

35. The mbira is a popular musical instrument of the Democratic Republic of the Congo and what other country?
Zimbabwe

36. The Jamkaran Mosque is just outside Qom, a city considered holy by many Shia Muslims, in what country?
Iran

37. Ataturk Olympic Stadium is in a Turkish city whose Ortakoy Mosque lies beside the Bosphorus Strait. Name this city.
Istanbul

38. The Stele Forest is a museum for stone sculptures located within a former Confucian temple in what Chinese city home to the Giant Wild Goose Pagoda?
Xi'an

39. Baklava is a sweet pastry made with layers of filo, chopped nuts, and usually honey or syrup. This dish is popular in the Eastern Mediterranean region, especially

in what country whose national sport is oil wrestling, of which a tournament is held annually in Edirne?
Turkey

40. German, Afrikaans, and Ovambo are the most spoken languages along with English in what country bordering the Portuguese-speaking country of Angola?
Namibia

41. The Mir Castle Complex is a cultural UNESCO World Heritage Site in what country?
Belarus

42. Shona and Ndebele are one of many languages spoken in what southern African country where bota and sadza are popular foods?
Zimbabwe

43. The Baitul Futuh Mosque is the largest mosque in Western Europe and is located in what country?
United Kingdom

44. The Tianning Temple is known for its pagoda, the tallest in the world. This Buddhist site is in Changzhou, in what Chinese province?
Jiangsu Province

45. Silesian is primarily spoken in Poland and what other Central European country?
Czech Republic

46. Boudhanath is a Buddhist stupa is located near Swayambhunath, an ancient site in what city in Nepal?
Kathmandu

47. Bukit Jalil National Stadium was the site of the 1998 Commonwealth Games in what city in Southeast Asia?
Kuala Lumpur

48. Pencak Silat is a martial art that originated in what island country home to the Prambanan Temple Complex?
Indonesia

49. The Meke is an art form originating in what country that uses a lovo, a type of earth oven, to cook foods such as palusami?
Fiji

50. The Airavatesvara Temple is an example of Dravidian architecture in the panchayat town of Darasuram in what South Indian state?
Tamil Nadu

51. The Meankieli dialects are spoken along the Torne River Valley and is native to Finland and what other country?
Sweden

52. The Guanyin of the South China Sea is a statue on what extinct volcano in China's Guangdong Province?
Mount Xiqiao

53. The cuatro is the national instrument of what South American country whose culture has been greatly influenced by Caribbean culture?
Venezuela

54. The Abbey of Santa Guistina is located in an Italian city on the Bacchiglione River and home to the Piazza dei Signori. Name this city.
Padua

55. Gagauz is a regional language of the Ukraine and what other country?
Moldova

56. Guangdong Olympic Center Stadium is the largest stadium by seating capacity in China and is in what major city home to the Chigang Pagoda?
Guangzhou

57. Kande Vihara is a Buddhist temple and an archaeological site in what island country?
Sri Lanka

58. Pahela Baishakh marks the first day of the calendar of what country home to the Kantajew Temple in Dinajpur?
Bangladesh

59. Stade Olympique de Rades is a stadium in Rades, in the Ben Arous Governorate of what country where Maghrebi Arabic is spoken?
Tunisia

60. Cristo de las Noas is a religious Christian statue in what Mexican city that is served by Francisco Sarabia International Airport?
Torreon

61. The Betawi language is native to what archipelagic country?
Indonesia

62. The Hausa language is now considered a lingua franca of many countries in West Africa and has national status in what country bordering Nigeria?
Niger

63. The Central Bikol language is spoken in the Bicol Region on what Philippine island?
Luzon

Economic Geography

1. What country is the world's largest producer of mango, ranks second worldwide in its labor force, and has a large information technology sector, especially in its southern region?
    India

2. The port of Tanjung Priok is located in what major city in Indonesia whose old port was known as Sunda Kelapa, situated on the Ciliwane River?
    Jakarta

3. This country is the world's largest producer of titanium and is home to over six hundred national parks. Name this country, which borders King Sound.
    Australia

4. What South American country that is part of the Southern Cone and borders Bolivia to the north is the world's third largest producer of soybean?
    Argentina

5. The world's largest producer of platinum borders the Atlantic Ocean on one side and the Indian Ocean on the other. Name this country.
    South Africa

6. What country bordering the Arabian Peninsula and has one of the smallest coastlines on the Persian Gulf is the world's third largest exporter of oil?
Iraq

7. The port city of Algeciras is one of the busiest container ports in Europe. Algeciras is located in the province of Cadiz, in the autonomous community of Andalusia in what country?
Spain

8. The world's largest producer of sugarcane is a BRIC country and has been one of the world's largest coffee producers coffee for the 150 years. Name this country.
Brazil

9. What country bordering the Gulf of Mexico and the Caribbean Sea is the world's second largest producer of bismuth?
Mexico

10. What Asian country bordering the Andaman Sea is the world's largest producer of sesame?
Myanmar

11. Copper is a major mineral resource in what country, the world's second largest producer of lithium after Australia?
Chile

12. What term describes a market where factor production services are sold?
Factor Market

13. What port, the busiest in Greece, has served Athens for a long time and is a seaport on the Mediterranean Sea?
Piraeus

14. What landlocked country bordering the Caspian Sea is the west is the world's largest producer of uranium, before Canada?
Kazakhstan

15. The port city of Taicang is located in Jiangsu Province and borders Nantong to the north across what river?
Yangtze River

16. The zloty is the currency of what country that is a member of the Schengen Area and has a major seaport at Szczecin, on the Bay of Pomerania?
Poland

17. Laem Chabang is located north of Pattaya and south of Chonburi. This port city is part of the Chonburi Province in what country?
Thailand

18. What country, home to the port of Santos, is the world's largest producer of oranges?
Brazil

19. What country is the world's second largest exporter of natural gas and is home to the industrial hub of Ras Laffan Industrial City?
Qatar

20. The price of a commodity can vary based on the factors of its amount and the need for that product. Name this economic model.
Supply and Demand

21. What country that straddles the Tropic of Cancer is the world's largest producer of pomegranate and has large ports on the Arabian Sea?
India

22. What is the official currency of Croatia?
Croatian kuna

23. What country is the world's largest producer of manganese and is known as the "Rainbow Nation" to describe its post-apartheid society?
South Africa

24. What is the economic model used by the People's Republic of China whose origins were in the Chinese economic reforms used by Deng Xiaoping?
Socialist Market Economy

25. What country has a free market mixed economy and is the world's second largest producer of jackfruit and jute after India?
Bangladesh

26. What country is the world's second largest producer of fluorite and is home to the Chicoasen Dam, the tallest in North America?
Mexico

27. What country is the world's second largest producer of hazelnuts and hosts the largest number of UNESCO World Heritage Sites in the world?
Italy

28. The tala is the currency of what Pacific island country?
Samoa

29. What Central Asian country is the world's third largest producer of apricot and has the world's fourth largest gold deposits?
Uzbekistan

30. Batam, an emerging transport hub in Indonesia, is part of the Indonesia-Malaysia-Singapore Growth Triangle and is home to Barelang Bridge in what Indonesian province?
Riau Islands Province

31. What East Asian country's period of rapid economic growth is known as the "Miracle on the Han River"?
South Korea

32. The Mekong-Ganga Corporation, consisting of six countries, was established in what Laotian city?
Vientiane

33. The Seat of Secretariat for the Caribbean Community is in what South American city nicknamed the "Garden City of the Caribbean"?
Georgetown

34. Name the bean that plays a large role in Cote d'Ivoire's and Ghana's economies.
Cocoa (Cacao)

35. What North American country is the world's largest producer of both strawberries and blueberries?
United States

36. One of the world's largest producers of silicon is known for its Gross National Happiness and has its national animal as the takin. Name this South Asian country.
Bhutan

37. What West African country has the ouguiya as its official currency and is home to the Chinguetti Oil Field?
Mauritania

38. What African country is ranked first in the world for its date production and is home to the headquarters of the Arab League?
Egypt

39. The Port of Keelung serves destinations to the Matsu Islands in what sovereign state?
Taiwan

40. The leu is the currency of what country?
Romania

41. The Eurasian Economic Union includes member states of northern Eurasia and has its largest city in what country?
Russia

42. The Organization for Economic Cooperation and Development, also known as OECD, originated to enforce the Marshall Plan and now has its headquarters at a château in what European city?
Paris

43. An increase in a good's or service's inflated market value over time measured as a percentage increased in real GDP, or gross domestic product, is known by what term?
Economic growth

44. The Laem Chabang container port is located in what country that is the world's largest producer of rubber?
Thailand

45. Name the regional bloc whose official languages are Spanish, Portuguese, and Guarani.
Mercosur

46. What rights determine how an economic good or resource is used, owned, or kept?
Property Rights

47. The European Union, established by the Maastricht Treaty in 1993, has its headquarters in what city?
Brussels

48. What landlocked East Asian country has the togrog as its official currency?
Mongolia

49. The Port of Felixstowe is the busiest container port in what country that is the world's fourth largest producer of wool and whose currency is the pound?
United Kingdom

50. The African Union is based in what city in Northeast Africa home to the headquarters of the United Nations Economic Commission for Africa?
Addis Ababa

51. The relationship between unemployment and production losses of a country is known by what law of economics?

Okun's Law

52. What term describes an increase in the price of economic goods and services of a country that results in a loss of currency value?
Inflation

53. The headquarters of the Asian Development Bank are in Mandaluyong, a city in what country that is the world's largest producer of nickel?
Philippines

54. A market where the government controls supply and demand, who enters the market, and prices on goods is known by what term?
Regulated Market (Controlled Market)

55. The gourde is the currency of what country whose economy was damaged by a major earthquake in 2010?
Haiti

56. The Andean Community is a customs union comprising four countries. This organization has its seat of secretariat in what Peruvian city that hosted the 2014 United Nations Climate Change Conference?
Lima

57. What country is the world's third largest producer of bentonite and has the largest economy on the Balkan Peninsula?

Greece

58. What term describes the monetary measure of any goods' or services' market value produced in a certain period of time?
Gross Domestic Product (GDP)

59. What country that is the world's third largest producer of pear is a regional power in the Southern Cone of South America and a member of the Group of Twenty?
Argentina

60. What country on the island of New Guinea has its currency as the kina?
Papua New Guinea

61. An economic system where transactions between parties do not experience government intervention is known by what French name?
Laissez-faire

62. What country, the world's third largest producer of okra, has the dinar as its official currency and is where OPEC was founded?
Iraq

63. The South Asian Association for Regional Cooperation launched the South Asian Free Trade Area in 2006 and was founded in 1985 in what city that is the seat of secretariat for the organization BIMSTEC?

Dhaka

64. What country is the world's second largest producer of tin and organized the Bandung Conference in 1955?
Indonesia

65. The Gulf Cooperation Council, whose current member states all consist of monarchies, has its headquarters in what Saudi Arabian city?
Riyadh

66. The Asia-Pacific Economic Cooperation forum has its headquarters in what country and city-state whose largest planning area is Bedok?
Singapore

67. What landlocked South Asian country is the world's third largest producer of ginger?
Nepal

68. The European Free Trade Area consists of four member nations, including what country that is a major producer of geothermal power and whose currency is the krona?
Iceland

69. Name the country whose currency is the lira.
Turkey

70. Places such as commercial companies and investment management firms make up a market where people can buy, sell, or exchange currency. Name this market.
    Foreign Exchange Market

71. Spindletop is a salt dome oil field that began the oil boom in what U.S. state in 1901, making the United States the world's largest producer of oil?
    Texas

Political Geography

1. Emmanuel Macron is the president of which Western European country that is the most popular tourist destination in the world?
   France

2. The Three Pagodas Pass is situated on the border between Thailand and what other country?
   Myanmar

3. Danilo Medina, the current president of the Dominican Republic, assumed office in 2012 and was born in Bohechio in what province?
   San Juan Province

4. The sea block of Ambalat off the northeastern coast of Borneo is the subject of a territorial dispute between Indonesia and what country?
   Malaysia

5. Viktor Orban, who lost in the 2002 and 2006 elections to the Socialist Party, was born in Szekesfehervar and is the current president of what country?
   Hungary

6. Name the legislative capital of Georgia, in the country's Imereti Region.
    Kutaisi

7. The Liancourt Rocks are located in the Sea of Japan and are disputed between Japan and what other country?
    South Korea

8. What form of government refers to a society ruled by the working class?
    Ergatocracy

9. Pedro Pablo Kuczynski is the current president of what country which gained independence from Spain in 1824?
    Peru

10. The David Gareja Monastery Complex is disputed between Georgia and what other country?
    Azerbaijan

11. What uninhabited area is located between Egypt and Sudan to the west of the Hala'ib Triangle and is claimed by neither country?
    Bir Tawil

12. Rukwanzi Island, located in Lake Albert, is disputed between Uganda and what other country?
    Democratic Republic of the Congo

13. Alpha Conde, who was born in Boke in the Boke Region, is the current president of what country?
Guinea

14. Mbabane, in the Mdimba Mountains, is the administrative capital of what country?
Swaziland

15. Kafia Kingi is a mineral-rich region disputed between Sudan and what bordering country?
South Sudan

16. Name the system of government where one person has supreme power.
Autocracy (Absolute Monarchy or Dictatorship)

17. Name the de facto temporary capital of the Sahrawi Arab Democratic Republic that succeeded Bir Lehlou.
Tifariti

18. Abu Musa Island, a strategic economic point in the eastern Persian Gulf, is disputed between Iran and what small country on the Arabian Peninsula?
United Arab Emirates

19. Horacio Cartes, a member of the Colorado Party, is the current president of what country whose government is a unitary presidential constitutional republic?
Paraguay

20. Hans Island, located in the Kennedy Channel of the Nares Strait, is claimed by Denmark for Greenland and disputed with what other country?
Canada

21. Name the Chilean city that is the site of the country's national legislature and the world's oldest Spanish newspaper.
Valparaiso

22. What city in Mozambique is known as the "City of Acacias" and the "Pearl of the Indian Ocean" and was formerly called Lourenco Marques before the country gained independence from Portugal?
Maputo

23. Yugoslavia, created in 1918 after World War I, eventually broke up into how many countries?
Seven

Historical

Geography/History

Happens

1. Confucius is believed to have been born at Mount Ni, near the Qufu in what Chinese province?
   Shandong Province

2. The Salt March, also known as the Dandi Satyagraha, was a nonviolent protest in 1930 in what Indian state with the archaeological sites of Dholavira and Lothal?
   Gujarat

3. What Asian country was formerly known as Siam and was once home to the Kingdom of Sukhothai , which lasted for 200 years?
   Thailand

4. The League of Nations, created in 1920 as a result of the Paris Peace Conference after World War I, had its headquarters in what Western European city known as the "Peace Capital"?
   Geneva

5. What city in Uttar Pradesh was founded by the Mughal Emperor Akbar in 1569 and is home to Buland Darwaza, the highest gateway in the world?
   Fatehpur Sikri

6. Italian East Africa was an area on the Horn of Africa, formed in 1936 after combining Italian Somaliland, Italian Eritrea, and what present day country?
   Ethiopia

7. What "Low Country" occupied the Democratic Republic of the Congo from 1908 to 1960 and controlled the Ruanda-Urundi territory around that same time period?
   Belgium

8. Name the only country in South America that was ruled by the Portuguese and at one time was part of a pluricontinental monarchy also consisting of the Kingdom of the Algarves.
   Brazil

9. What city on the island of Hispaniola was once known as Cap-Francais and is home to the Citadelle Laferriere?
   Cap-Haitien

10. What country was formerly known as Kampuchea to the Khmer People and is home to the Bakong Temple Mountain in the ancient city of Hariharalaya?
    Cambodia

11. The New Netherland colony occupied much of the East Coast in the 17th century, extending from Cape Cod to what peninsula that partially became an island after the construction of the Chesapeake and Delaware Canal?
Delmarva Peninsula

12. Forty-eight countries signed what treaty in the September of 1851 that ended Japan's position as an imperial power?
Treaty of San Francisco (Peace Treaty with Japan)

13. The Western European Union was an international organization that enforced the 1954 Modified Treaty of Brussels during what war consisting of small proxy wars?
Cold War

14. Yorubaland, the cultural region of the Yoruba people, is now divided among the countries of Togo, Benin, and what country that is now home to the once secessionist state of the Republic of Biafra?
Nigeria

15. Name the treaty signed that divided up land between Portugal and the Crown of Castile, now part of present day Spain, that divided up land between both countries.
Treaty of Tordesillas

16. Name the famous canal that was opened across an isthmus in 1914 by the United States.

Panama Canal

17. Member countries of the Warsaw Pact had their headquarters in what major city that was once part of the Soviet Union?
Moscow

18. The Kansas-Nebraska Act of 1854 created the territories of Kansas and Nebraska and repealed what legislation passed in 1820?
Missouri Compromise

19. The capture of Saigon, now known as Ho Chi Minh City, ended what war on the Indochinese peninsula?
Vietnam War

20. The historical capital of Moravia was Brno, which is the second largest city in what country?
Czech Republic

21. The Zapotec Civilization could be found in the Oaxaca Valley in modern day Mexico, and left evidence of their community at what famous archaeological site near Oaxaca City?
Monte Alban

22. The Ashanti Empire, whose capital was at Kumasi, was built around Lake Bosumtwi in what present day West African country?
Ghana

23. The Adams-Onis Treaty of 1819 ceded Florida to the United States and defined the border between the United States and what colonial Spanish territory spread out north of the Isthmus of Panama?
New Spain

24. In 1946, what Asian archipelagic country gained independence from the United States?
Philippines

25. Name the boundary that separated two parts of Europe from World War II to the Cold War.
Iron Curtain

26. Name the largest empire in Pre-Columbian America, whose people worshipped the sun god Inti.
Incan Empire

27. The Battle of Gettysburg occurred during the Civil War in 1863, in what state?
Pennsylvania

28. Ubangi-Shari was a former French colony that received its independence in 1960 and is now known by what name?
Central African Republic

29. What treaty led to the breakup of the Ottoman Empire and was signed by the Central Powers after World War I in 1920?
Treaty of Sevres

30. Ayutthaya, the former capital of the Phra Nakhon Si Ayutthaya, was once one of the largest cities in the world and was sometimes known as the "Venice of the East". This city is named after the ancient Indian city of Ayodhya and is located in what country?
Thailand

31. During the Battle of Remagen, American forces captured Ludendorff Bridge over the Rhine River in what present day country?
Germany

32. What act was passed during Andrew Jackson's presidency and relocated thousands of Cherokee to Oklahoma on a journey called the "Trail of Tears"?
Indian Removal Act

33. Failaka Island was occupied by traders from the Sumerian city of Ur around 2000 B.C. This island is now in what country?
Kuwait

34. Andijan was an important city on the Silk Road and widely regarded as the birthplace of Babur. This country is located in the Fergana Valley of what country?

Uzbekistan

35. Name the treaty that brought an end to World War I and was signed in 1919 in Paris, France.
Treaty of Versailles

36. The Viceroyalty of Peru was established in 1542 in Lima by what colonial power?
Spain

37. The Buganda Kingdom was the kingdom of the Ganda people in the present day Central Region of what country?
Uganda

38. The dual monarchy of Austria and Hungary was created by the Austro-Hungarian Compromise of what year?
1867

39. The Yalta Conference took place in 1945 at the Livadia Palace in what peninsular region?
Crimea

40. What agreement signed in 1917 led to the creation of the Kingdom of Yugoslavia in 1918?
Corfu Declaration

Miscellaneous

River Confluences

Instead of asking about what confluence the city is at, these questions ask about what city is at the confluence. If you have never heard of the confluence, it is a good idea to listen to language patterns in the river names. If this doesn't work, give an answer such that the city you name has the same name as one of the rivers.

1. What U.S. city is located at the confluence of the Ashley and Cooper Rivers?
   Charleston

2. What city in Laos is located at the confluence of the Nam Khan and Mekong Rivers?
   Luang Prabang

3. What city in Russia is located at the confluence of the Volga and Oka Rivers?
   Nizhny Novgorod

4. What U.S. city is located at the confluence of the St. Johns and Trout Rivers?
   Jacksonville

5. What city in Canada is located at the confluence of the Ottawa, Gatineau, and Rideau Rivers?
    Ottawa

6. What U.S. city is located at the confluence of the Sacramento and American Rivers?
    Sacramento

7. What city in India is located at the confluence of the Ganges, Son, Gandaki, and Punpun Rivers?
    Patna

8. What city in Brazil is located at the confluence of the Capibaribe and Beribe Rivers?
    Recife

9. What Slovakian city is located at the confluence of the Danube and Vah Rivers?
    Komarno

10. What U.S. city is located at the confluence of the Allegheny and Monongahela Rivers?
    Pittsburgh

11. What Pakistani city is located at the confluence of the Indus and Shigar Rivers?
    Skardu

12. What U.S. city is located at the confluence of the Willamette and McKenzie Rivers?

Eugene

13. What city in Switzerland is located at the confluence of the Arve and Rhone Rivers?
Geneva

14. What U.S. city is located at the confluence of the Great Miami and Mad Rivers?
Dayton

15. What city in India is located at the confluence of the Yamuna and Ganges Rivers?
Allahabad

16. What city in Germany is located at the confluence of the Spree and Havel Rivers?
Berlin

17. What U.S. city is located at the confluence of the Yukon and Tanana Rivers?
Tanana

18. What Russian city is located at the confluence of the Kostroma and Volga Rivers?
Kostroma

19. What U.S. city is located at the confluence of the Arkansas and Little Arkansas Rivers?
Wichita

20. What city in Canada is located at the confluence of the Red and Assiniboine Rivers?
Winnipeg

21. What U.S. city is located near the confluence of the Hudson and Mohawk Rivers?
Albany

22. What city in Canada is located at the confluence of the Bow and Elbow Rivers?
Calgary

23. What Indian city is located at the confluence of the Ghaghara and Ganges Rivers?
Ballia

24. What Russian city is located at the confluence of the Yenisei and Kacha Rivers?
Krasnoyarsk

25. What German city is located at the confluence of the Inn, Ilz, and Danube Rivers?
Passau

26. What city in Pakistan is located at the confluence of the Indus and Kabul Rivers?
Attock

27. What U.S. city is located at the confluence of the Des Moines and Raccoon Rivers?

Des Moines

28. What Belgian city is located at the confluence of the Scheldt and Leie Rivers?
Ghent

29. What city in Canada is located at the confluence of the St. Lawrence and Ottawa Rivers?
Montreal

30. What city in Nigeria is located at the confluence of the Niger and Benue Rivers?
Lokoja

31. What Croatian city is located at the confluence of the Vuka and Danube Rivers?
Vukovar

32. What city in Russia is located at the confluence of the Sheksna and Volga Rivers?
Cherepovets

33. What Venezuelan city is located at the confluence of the Orinoco and Caroni Rivers?
Ciudad Guayana

34. What city in Ukraine is located at the confluence of the Dnieper and Samara Rivers?
Dnipropetrovsk

35. What city in Laos is located at the confluence of the Xe and Mekong Rivers?
Pakxe

36. What U.S. city is located at the confluence of the Delaware and Schuylkill Rivers?
Philadelphia

37. What city in the United Kingdom is located at the confluence of the Thames and Cherwell Rivers?
Oxford

38. What U.S. city is located at the confluence of the Shatucket and Yantic Rivers?
Norwich

39. What Lithuanian city is located at the confluence of the Vilnia and Neris Rivers?
Vilnius

40. What Chinese city is located at the confluence of the Han and Yangtze Rivers?
Wuhan

41. What city in Russia is located at the confluence of the Akhtuba and Volga Rivers?
Volzhsky

42. What German city is located at the confluence of the Brigach and Berg Rivers?

Donaueschingen

43. What Sudanese city is located at the confluence of the Nile and Atbara Rivers?
Atbara

44. What Chinese city is located at the confluence of the Yangtze and Jialing Rivers?
Chongqing

45. What city in Argentina is located at the confluence of the Salado and Parana Rivers?
Santa Fe

46. What Swiss city is located at the confluence of the Aare, Limmat, and Reuss Rivers?
Brugg

47. What U.S. city is located near the confluence of the Schuylkill River and Tulpehocken Creek?
Reading

48. What U.S. city is located at the confluence of the Columbia and Willamette Rivers?
Portland

49. What two U.S. cities are located at the confluence of the Mississippi and Minnesota Rivers?
Saint Paul and Minneapolis

50. What city in Russia is located at the confluence of the Yenisei and Abakan Rivers?
Abakan

51. What Chinese city is located at the confluence of the Luo and Yellow Rivers?
Luoyang

52. What Peruvian city is located near the confluence of the Nanay and Amazon Rivers?
Iquitos

53. What U.S. city is located at the confluence of the Kansas and Missouri Rivers?
Kansas City

54. What Russian city is located at the confluence of the Kazanka and Volga Rivers?
Kazan

55. What city in Slovakia is located at the confluence of the Danube and Hron Rivers?
Sturovo

56. What city in Mongolia is located at the confluence of the Tuul and Selbe Rivers?
Ulaanbaatar

57. What U.S. city is located at the confluence of the Wolf and Mississippi Rivers?

Memphis

58. What city in France is located at the confluence of the Rhone and Saone Rivers?
Lyon

59. What Slovenian city is located at the confluence of the Kokra and Sava Rivers?
Kranj

60. What city in Germany is located at the confluence of the Rhine and Neckar Rivers?
Mannheim

61. What Chinese city is located at the confluence of the Jinsha and Batang Rivers?
Yushu

62. What city in Russia is located at the confluence of the Volga and Tvertsa Rivers?
Tver

63. What U.S. city is located at the confluence of the Mississippi and Missouri Rivers?
St. Louis

64. What Russian city is located at the confluence of the Volga and Kotorosl Rivers?
Yaroslavl

65. What city in Romania is located at the confluence of the
    Siret, Prut, and Danube Rivers?
    Galati

66. What city in India is located at the confluence of the
    Dhauliganga and Alaknanda Rivers?
    Vishnuprayag

67. What U.S. city is located near the confluence of the
    Potomac and Anacostia Rivers?
    Washington D.C.

68. What English city is located at the confluence of the
    Severn and Avon Rivers?
    Tewkesbury

69. What German city is located at the confluence of the
    New Rhine and Old Rhine Rivers?
    Leiden

70. What Serbian city is located at the confluence of the
    Danube and Timis Rivers?
    Pancevo

71. What U.S. city is located at the confluence of the Clark
    Fork, Bitterroot, and Blackfoot Rivers?
    Missoula

72. What Austrian city is located at the confluence of the
    Danube and Krems Rivers?

Krems

73. What city in Pakistan is located at the confluence of the Chenab and Jhelum Rivers?
Jhang

74. What city in Sudan is located at the confluence of the Blue Nile and White Nile Rivers?
Khartoum

75. What city in Belarus is located at the confluence of the Pripyat and Strumen Rivers?
Pinsk

76. What German city is located at the confluence of the Danube, Naab, and Regen Rivers?
Regensburg

77. What Russian city is located at the confluence of the Lena and Vitim Rivers?
Vitim

78. What German city is located at the confluence of the Mosel and Rhine Rivers?
Koblenz

79. What city in China is located at the confluence of the Rong and Long Rivers?
Fengshan

80. What Canadian city is located near the confluence of the Wrigley and Mackenzie Rivers?
    Wrigley

81. What city in Russia is located at the confluence of the Irtysh and Om Rivers?
    Omsk

82. What U.S. city is located at the confluence of the Klondike and Yukon Rivers?
    Dawson City

83. What city in Slovakia is located near the confluence of the Danube and Morava Rivers?
    Bratislava

84. What city in Botswana is located at the confluence of the Tata and Inchwe Rivers?
    Francistown

85. What city in Russia is located at the confluence of the Samara and Volga Rivers?
    Samara

86. What city in Canada is located at the confluence of the Kootenay and Columbia Rivers?
    Castlegar

87. What Iraqi city is located at the confluence of the Tigris and Euphrates Rivers?

Al-Qurnah

88. What French city is located at the confluence of the Durance and Rhone Rivers?
Avignon

89. What U.S. city is located at the confluence of the Scioto and Olentangy Rivers?
Columbus

90. What Indian city is located at the confluence of the Alaknanda and Bhagirathi Rivers?
Devprayag

91. What U.S. city is located at the confluence of the Yukon and Tanana Rivers?
Fairbanks

92. What city in Romania is located at the confluence of the Danube and Olt Rivers?
Turnu Magurele

93. What Welsh city is located at the confluence of the Seven and Taff Rivers?
Cardiff

94. What city in the United States is located at the confluence of the South Platte River and Cherry Creek?
Denver

95. What city in Canada is located at the confluence of the Mackenzie and Great Bear Rivers?
Tulita

96. What city in Germany is located at the confluence of the Elbe, Alster, and Bille Rivers?
Hamburg

97. What Chinese city is located at the confluence of the Second Songhua and Amur Rivers?
Tongjiang

98. What Congolese city is located at the confluence of the Kwango and Kwilu Rivers?
Bandundu

99. What U.S. city is located at the confluence of the Potomac and Shenandoah Rivers?
Harpers Ferry

100. What Malaysian city is located at the confluence of the Klang and Gombak Rivers?
Kuala Lumpur

101. What U.S. city is located at the confluence of the Clearwater and Snake Rivers?
Lewiston

102. What Brazilian city is located at the confluence of the Tocantins and Araguaia Rivers?

Maraba

103. What Taiwanese city is located at the confluence of the Xindian and Dahan Rivers?
Taipei

104. What Ukrainian city is located at the confluence of the Inhulets and Saksahan Rivers?
Kryvyi Rih

105. What city in Lithuania is located at the confluence of the Nemunas and Neris Rivers?
Kaunas

106. What Indian city is located at the confluence of the Chambal and Yamuna Rivers?
Etawah

107. What city in Germany is located at the confluence of the Main and Rhine Rivers?
Mainz

108. What Chinese city is located at the confluence of the Gui and Xun Rivers?
Wuzhou

109. What city in the Democratic Republic of the Congo is located at the confluence of the Congo and Tshuapa Rivers?
Mbandaka

110. What U.S. city is located at the confluence of the Crow and Dry Creeks?
Cheyenne

111. What city in Croatia is located at the confluence of the Kupa, Sava, and Odra Rivers?
Sisak

112. What city in India is located at the confluence of the Jalangi and Bhagirathi Rivers?
Nabadwip

113. What German city is located at the confluence of the Altmuhl and Danube Rivers?
Kelheim

114. What city in Uzbekistan is located at the confluence of the Kara Darya and Naryn Rivers?
Namangan

115. What city in China is located at the confluence of the Wu and Zhen Rivers?
Shaoguan

116. What group of U.S. cities are located at the confluence of the Rock and Mississippi Rivers?
Quad Cities

117. What city in Romania is located near the confluence of the Bistrita and Siret Rivers?
Bacau

118. What U.S. city is located at the confluence of the St. Croix and Mississippi Rivers?
Prescott

119. What Canadian village is located at the confluence of the Mackenzie and Liard Rivers?
Fort Simpson

120. What town in Russia is located at the confluence of the Lena and Kirenga Rivers?
Kirensk

121. What city in Germany is located at the confluence of the Blau, Iller, and Danube Rivers?
Ulm

122. What city in the Democratic Republic of the Congo is located at the confluence of the Congo, Tshopo, and Lindi Rivers?
Kisangani

123. What city in the United States is located at the confluence of the Smoky Hill and Republican Rivers?
Junction City

124. What Argentinian city is located near the confluence of the Parana and Paraguay Rivers?
Corrientes

125. What Russian city is located at the confluence of the Northern Dvina and Vychegda Rivers?
Kotlas

126. What city in France is located at the confluence of the Loire and Erdre Rivers?
Nantes

127. What city in South Africa is located near the confluence of the Orange and Vaal Rivers?
Kimberly (Douglas is also acceptable)

128. What Austrian city is located near the confluence of the Mur and Sulm Rivers?
Leibnitz

129. What city in Australia is located near the confluence of the Maribyrnong and Yarra Rivers?
Melbourne

130. What Malaysian city is located at the confluence of the Perak and Kangsar Rivers?
Kuala Kangsar

131. What U.S. city is located near the confluence of the Iowa and English Rivers?

Iowa City

132. What city in Russia is located at the confluence of the Barnaulka and Ob Rivers?
Barnaul

133. What Filipino city is located near the confluence of the Tarlac and Agno Rivers?
Bayambang

134. What Ecuadorian city is located at the confluence of the Daule and Babahoyo Rivers?
Guayaquil

135. What city in the United States is located near the confluence of the Smoky Hill and Saline Rivers?
Salina

136. What Chinese city is located near the confluence of the Zuo and Yuo Rivers?
Nanning

137. What Paraguayan city is located near the confluence of the Parana and Paraguay Rivers?
Pilar

138. What Russian city is located at the confluence of the Amur and Ussuri Rivers?
Khabarovsk

139. What city in India is located at the confluence of the Damodar and Barakar Rivers?
Dishergarh

140. What Iranian city is located at the confluence of the Aji and Quri Rivers?
Tabriz

141. What U.S. city is located at the confluence of the Merrimack and Soucook Rivers?
Concord

142. What Luxembourgish city is located at the confluence of the Alzette and Petrusse Rivers?
Luxembourg City

143. What Argentinian city is located near the confluence of the Iguazu and Parana Rivers?
Puerto Iguazu

144. What city in India is located at the confluence of the Girna and Mausam Rivers?
Malegaon

145. What Canadian city is located at the confluence of the Athabasca, Clearwater, Hangingstone, and Horse Rivers?
Fort McMurray

146. What Chinese city is located at the confluence of the Pearl and Xi Rivers?

Zhaoqing

147. What U.S. city is located at the confluence of the Driftwood and Flatrock Rivers?
Columbus (in the state of Indiana)

148. What city in Australia is located at the confluence of the Barwon and Namoi Rivers?
Walgett

149. What Indian city is located near the confluence of the Ganges and Sone Rivers?
Danapur

150. What city in the United States is located at the confluence of the Salmon and Lemhi Rivers?
Salmon

151. What Australian city is located at the confluence of the Richmond and Wilson Rivers?
Coraki

152. What city in Chad is located near the confluence of the Chari and Logone Rivers?
N'Djamena

153. What French city is located at the confluence of the Loir, Mayenne, and Sarthe Rivers?
Angers

154. What Paraguayan city is located near the confluence of the Iguazu and Parana Rivers?
Ciudad del Este

155. What Canadian city is located at the confluence of the St. Lawrence and St. Maurice Rivers?
Trois-Rivières

156. What Russian city is located at the confluence of the Belaya and Ufa Rivers?
Ufa

157. What Ecuadorian city is located near the confluence of the Cutuchi and Alaquez Rivers?
Latacunga

158. What city in Spain is located at the confluence of the Tagus and Jarama Rivers?
Aranjuez

159. What city in the Democratic Republic of the Congo is located at the confluence of the Dungu and Kibali Rivers?
Dungu

160. What U.S. city is located at the confluence of the Saluda and Broad Rivers?
Columbia

161. What Belarusian city is located at the confluence of the Bug and Mukhavets Rivers?

Brest

162. What South African city is located at the confluence of the Vaal and Klip Rivers?
Vereenigeng

163. What city in Bangladesh is located near the confluence of the Louhajang and Jamuna Rivers?
Tangail

164. What Romanian city is located near the confluence of the Danube and Olt Rivers?
Turnu Magurele

165. What Russian city is located at the confluence of the Uda and Selenge Rivers?
Ulan-Ude

166. What city in India is located at the confluence of the Krishna and Koyna Rivers?
Karad

167. What Belgian city is located at the confluence of the Dender and Scheldt Rivers?
Dendermonde

168. What Brazilian city is located near the confluence of the Iguazu and Parana Rivers?
Foz do Iguacu

169. What city in Myanmar is located at the confluence of the Yangon and Bago Rivers?
Yangon

170. What Indonesian city is located near the confluence of the Barito and Martapura Rivers?
Banjarmasin

171. What German city is located at the confluence of the Fulda and Werra Rivers?
Hannoversch Munden (Hann Munden)

172. What Russian city is located at the confluence of the Lena and Kuta Rivers?
Ust-Kut

173. What city in Canada is located at the confluence of the Bécancour and St. Lawrence Rivers?
Bécancour

174. What Russian city is located at the confluence of the Amur and Zeya Rivers?
Blagoveshchensk

175. What city in Kosovo is located at the confluence of the Ibar and Sitnica Rivers?
Mitrovica

176. What Spanish city is located at the confluence of the Pisuerga and Esgueva Rivers?

Valladolid

177. What city in Poland is located at the confluence of the
Vistula and Sola Rivers?
Oswięcim

178. What city in Switzerland is located at the confluence of
the Limmat and Sihl Rivers?
Zurich

179. What German city is located at the confluence of the
White Elster, Pleisse, and Parthe Rivers?
Leipzig

180. What Chilean city is located at the confluence of the
Rahue and Damas Rivers?
Osorno

181. What Czech city is located at the confluence of the
Svitava and Svratka Rivers?
Brno

182. What city in Serbia is located near the confluence of the
Ibar and Western Morava Rivers?
Kraljevo

183. What French city is located at the confluence of the
Moselle and Seille Rivers?
Metz

184. What city in Montenegro is located at the confluence of the Moraca and Ribnica Rivers?
Podgorica

185. What city in South Korea is located at the confluence of the Geumho and Nakdong Rivers?
Daegu

186. What Chinese city is located at the confluence of the Yangtze and Min Rivers?
Yibin

187. What Indian city is located at the confluence of the Aji and Nyari Rivers?
Rajkot

Cities on Rivers

1. Rishikesh, known as the "Yoga Capital of the World", is one of the holiest places in India for Hindus. This pilgrimage site, one of the most populous cities in Uttarakhand, is on what river?
Ganges River

2. What city in Japan is located on the Sumida River?
Tokyo

3. Name the city in Vietnam that lies on the Red River in Southeast Asia.
    Hanoi

4. The capital of an East African country bordering Ethiopia to the north is located on a river of the same name. Name this capital city.
    Nairobi

5. What city in southeastern Brazil lies on the eastern bank of the Guaiba River?
    Porto Alegre

6. The Guaire River, which empties out into the Tuy River, flows through what major city in Venezuela?
    Caracas

7. What city that lies on the Seine River is home to the Arc de Triomphe?
    Paris

8. What city in Ukraine is the third largest in Eastern Europe and is on the Dnieper River?
    Kiev

9. What city in West Bengal that is the commercial, educational, and cultural center of Eastern India is located on the eastern bank of the Hooghly River?
    Kolkata

10. What city in Central Vietnam is located on the banks of the Perfume River and was formerly the imperial capital of Vietnam?
Hue

11. What city, the southernmost in Colombia, is a major port city on the Amazon River?
Leticia

12. What city in China, which is home to the Qutang Gorge, is located on the Yangtze River?
Chongqing

13. What city in Armenia that is situated on the Ararat Plain is home to the Karen Demirchyan Complex and is located on the Hrazdan River?
Yerevan

14. What city in the Estuaire Province of Gabon is located on the Komo River?
Libreville

15. What major port city and economic hub is home to Igbo-speaking inhabitants in the Anambra State, on the Niger River?
Onitsha

16. What city home to the National Pantheon of the Heroes and is on the Paraguay River, near its confluence with the Pilcomayo River?

Asuncion

17. What major Iraqi city is located near the ruins of Ur and Larsa, is the capital of the Dhi Qar Governorate, and is situated on the Euphrates River?
Nasiriyah

18. What city is at the point on the Lower Volga River meets the Caspian Sea in Russia?
Astrakhan

19. What city is the capital of the Brazilian state of Rondonia and is on the eastern shore of the Madeira River?
Porto Velho

20. The Deh Cho Bridge is near Fort Providence and the Great Slave Lake, in the Northwest Territories. Fort Providence is on what major North American river?
Mackenzie River

21. Khortytsya Island is within the city limits of Zaporizhia, in Ukraine, on what major river?
Dnieper River

22. Vila Nova de Gaia is located in Portugal's Norte Region, on what river that originates in the country's Soria Province?
Portugal

23. What city in northeastern Thailand lies on the Mekong River, near the country's first Thai-Lao Friendship Bridge?
Nong Khai

24. The Chateau of the Dukes of Bretagne is a famous castle now in what city that is one of the most populous in France and is also situated on the Loire River?
Nantes

25. The JK Temple is located in Kanpur, a major city in Uttar Pradesh on what large Indian river?
Ganges River

Tallest Buildings and Structures

These questions will help you watch out for questions related to the tallest towers, buildings, structures, and skyscrapers in the world, especially in the USGO Finals and National Geographic Bee.

1. The Lotte World Tower is located in the capital city of Seoul, in South Korea, near what river?
Han River

2. The Al Hamra Tower is located in what city in the Al Asimah Governorate of Kuwait?
Kuwait City

3. The Telekom Tower is a skyscraper in what major Southeast Asian city bordering the Titiwangsa Mountains to the east?
    Kuala Lumpur

4. The Federation Tower is located in the Moscow International Business Center in what country?
    Russia

5. The Aon Center is the third tallest building in Chicago and is located in what central business district?
    Chicago Loop

6. The Commerzbank Tower can be found in the Innenstadt District of what German city on the Main River?
    Frankfurt

7. The Baiyoke Tower II is the tallest hotel in Southeast Asia and is located in what country?
    Thailand

8. The Greenland Puli Center is a skyscraper in Jinan in what Chinese province?
    Shandong Province

9. Q1 is a skyscraper in the Surfers Paradise suburb in the Gold Coast of what Australian state?
    Queensland

10. The Keangnam Hanoi Landmark Tower is in the Tu' Liem District of what Vietnamese city?
Hanoi

11. The Ryugyong Hotel is an unopened and pyramid-shaped tower located in what North Korean city?
Pyongyang

12. Aspire Tower is located near Khalifa International Stadium in what Persian Gulf country?
Qatar

13. The Kingdom Center is a skyscraper home to a shopping mall in Riyadh in what country?
Saudi Arabia

14. The Gran Torre Santiago is the tallest building in Latin America and is located in what country?
Chile

15. The Abenobashi Terminal Building is a commercial building and the tallest structure in Japan. This skyscraper is located in what city?
Osaka

16. Name the second tallest building in Dubai, after the Burj Khalifa, in the city's Marina District.
Princess Tower

17. The Commerzbank Tower is the tallest building in Germany and can be found in the banking district of what city?
Frankfurt

18. The Shimao International Plaza is located in the Huangpu District of what city?
Shanghai

Tunnels and Bridges

The building of tunnels and bridges is something that shows up a number of times when studying Current Events at the National Geographic Bee, so these questions will help you get ready.

1. The Paijianne Water Tunnel is the second longest tunnel in the world, connecting Lake Paijianne to what southern Finnish city?
Helsinki

2. The Zelivka Water Tunnel is located in Central Bohemia in what country?
Czech Republic

3. The Great Seto Bridge connects the Okayama and Kagawa Prefectures in Japan over what sea?
Seto Inland Sea (Inland Sea)

4. The Emisor Oriente Tunnel runs from Mexico City to what state bordering Queretaro to the west and Veracruz to the north?
    Hidalgo

5. The Wuhu Yangtze River Bridge crosses the Yangtze River at Wuhu in what province?
    Anhui Province

6. The Penang Bridge crosses over the Penang Strait and is located in what country?
    Malaysia

7. The Orange-Fish River Tunnel is an irrigation tunnel in the Eastern Cape Province of what country?
    South Africa

8. Name the longest bridge in Europe, built over the Tagus River in Lisbon, Portugal.
    Vasco da Gama Bridge

9. The Third Mainland Bridge crosses over Lagos Lagoon and connects what island to the Nigerian mainland?
    Lagos Island

10. The Rio-Niteroi Bridge in Rio de Janeiro, Brazil, passes over what bay?
    Guanabara Bay

11. The Dahuofang Water Tunnel provides water to cities such as Shenyang and Fushan in what Chinese province?
Liaoning Province

12. The General Rafael Urdaneta Bridge crosses Lake Maracaibo in what South American country?
Venezuela

13. The Jubilee Parkway crosses Mobile Bay in what U.S. state?
Alabama

14. The Bolmen Water Tunnel exits Lake Bolmen and reaches the Scania Province of what country?
Sweden

15. The Seikan Tunnel, which connects the Japanese islands of Honshu and Hokkaido, lies beneath what strait?
Tsugaru Strait

16. The Lotschberg Base Tunnel is a railway tunnel going through the Bernese Alps in what country?
Switzerland

17. Confederation Bridge crosses the Northumberland Strait linking what Canadian province to New Brunswick?
Prince Edward Island

18. Kamchiq Tunnel, the longest tunnel in Central Asia, is located along the Angren-Pap Railway in what country?

Uzbekistan

19. The Bang Na Expressway crosses the Bang Pakong River in what country's Prachinburi Province?
Thailand

20. The Oresund Bridge connects Sweden and what other country across the Oresund Strait?
Denmark

21. Gwangandaegyo, or the Diamond Bridge, is located in Busan, a major city in what country?
South Korea

22. The Thanlwin Bridge connects the cities of Mawlamyine and Mottama in what country?
Myanmar

23. The Libertador General San Martin Bridge connects the Rio Negro Department of Uruguay with the Entre Rios Province of what other country?
Argentina

24. The President Bridge, the second longest bridge in Russia, crosses over Ulyanovsk Oblast on what river?
Volga River

25. Name the longest river bridge in India, which connects the cities of Patna and Hajipur in the state of Bihar.
Mahatma Gandhi Setu

Twin Cities

1. The metropolitan area of San Diego borders what Mexican city to its south?
    Tijuana

2. Valga in Estonia is a border town with Valka in what country?
    Latvia

3. Komarno, a town at the confluence of the Danube and Vah Rivers, is a border city with what Hungarian city?
    Komarom

4. Strasbourg is a border city on the Rhine River adjacent to what German city in Baden-Wurttemberg?
    Kehl

5. Hampton Roads is a metropolitan region consisting of multiple bordering cities in what U.S. state?
    Virginia

6. Albury, on the northern part of the Murray River, is a border city with Wodonga in what country?
    Australia

7. Dipolog and Dapitan are border cities in the province of Zamboanga del Norte in what country?

Philippines

8. Asansol and Durgapur are border cities in what Indian state?
West Bengal

9. Chui, Brazil, is a border town with Chuy in what country whose capital is Montevideo?
Uruguay

10. Valparaiso, home of South America's oldest stock exchange, is a border city with what other Chilean city?
Vina del Mar

11. The border town of Phuntsholing in Bhutan is adjacent to what Indian city on the banks of the Torsa River?
Jaigaon

12. Name the southernmost city in peninsular Malaysia, which is a border town with Woodlands in Singapore.
Johor Bahru

13. El Paso, in the United States, is a border city with Ciudad Juarez, which is located in what Mexican state?
Chihuahua

14. Texarkana is a twin city with itself, in Texas and what other U.S. state?
Arkansas

15. Blagoveshchensk, in Russia, is a border city with Heihe, a city in what country?
China

Geographic Extremes

This round is crucial in the National Round of the National Geographic Bee, the USGO/IGB National Qualifying Exam, and the NSF Senior GeoBee Phase 1 Round. While this round focuses on geographic extremes, it includes manmade extremes as well.

1. The longest navigable road or highway in the world spans two continents and only has one break, at the Darien Gap. Name this road.
Pan-American Highway

2. One of the world's largest plains lies between the Ural Mountains and the Yenisei River Valley. Name this plain.
West Siberian Plain

3. The tallest building in the world can be found in Dubai in what country?
United Arab Emirates

4. Name the largest country on the largest peninsula in the world, bordering the Red Sea and the Persian Gulf.
Saudi Arabia

5. Name the westernmost point in Africa, an island that is the largest of Cape Verde's Barlavento Islands.
Santo Antao

6. One of the highest astronomical observatories in the world is the Sphinx Observatory above the Jungfraujoch, in the Bernese Alps in what country?
Switzerland

7. Name the highest point on the North American Continental Divide and in the Front Range, in Colorado.
Grays Peak

8. The world's shortest river as recognized by the Guinness Book of World Records is the Roe River, in Great Falls in what U.S. state?
Montana

9. Kidd Mine is the world's deepest copper and zinc mine and is located in Timmins, near the Mattagami River in what Canadian province?
Ontario

10. The world's longest bridge is the Danyang-Kunshan Grand Bridge, in the Jiangsu Province of what country?
China

11. What U.S. state is considered the country's westernmost, but extends across the 180th meridian, making it also the easternmost state?

Alaska

12. The world's flattest place is described as a famous salt flat in Bolivia. Name this salt flat.
Salar de Uyuni

13. When Greenland is not included what is the easternmost point in North America, on the Avalon Peninsula in Newfoundland, Canada?
Cape Spear

14. Name the Colombian department described as the world's rainiest lowland and the only administrative division in Colombia bordering both the Pacific and Atlantic Oceans.
Choco Department

15. What peninsula is the northernmost in mainland Canada, on the Boothia Peninsula in Nunavut?
Murchison Promontory

16. The deepest open-pit mine in the world, Tagebau Hambach, is in what German state whose capital is Dusseldorf?
North Rhine-Westphalia

17. A limestone karst river cave known as Lamprechtsofen is one of the deepest caves in the world and in Europe. This cave is in the Leogang Mountains in what country?
Austria

18. Monchique Islet is located off the coast of Flores Island and is the westernmost point in Europe. This islet is administered by Portugal and can be found in what archipelago?
    Azores Islands

19. The westernmost point in Asia is at Cape Baba in what country?
    Turkey

20. The northernmost point on the northernmost peninsula in Norway is the northernmost mainland point in Europe. Name this cape.
    Cape Nordkinn

21. The southernmost point in all of North America is an island in a national park that belongs to Costa Rica. Name this island.
    Cocos Island

22. The most populous landlocked country in the world was cut off from its access to the Red Sea after Eritrea's independence. Name this country.
    Ethiopia

23. Earth's greatest vertical drop is located at a mountain in Auyuittuq National Park on Baffin Island. Name this mountain.
    Mount Thor

24. What town, located along the Indigirka River in Russia, is the coldest inhabited place on Earth?
Oymyakon

25. The westernmost point in mainland Africa is Pointe des Almadies, located on what peninsula in Senegal?
Cap Vert Peninsula

26. What lake shared by the United States and Canada has the greatest volume in the Western Hemisphere?
Lake Superior

27. What is the highest point in Central America, in the San Marcos Department of Guatemala?
Volcan Tajumulco

28. Krubera Cave, the world's deepest cave, is located in what massif of the Gagra Mountains in Abkhazia?
Arabika Massif

29. The lowest point in North America is in an endorheic basin in Death Valley National Park, California. Name this basin.
Badwater Basin

30. The largest freshwater lake island in the world belongs to Canada and can be found in Lake Huron. Name this lake.
Lake Manitoulin

31. Name the point that is the southernmost point on the Iberian Peninsula and in Europe.
   Punta de Tarifa (Tarifa Point)

32. The easternmost point in Asia is at Big Diomede Island in Russia, located in what strait?
   Bering Strait

33. Commonwealth Bay, one of the windiest places on Earth, is situated between Point Alden and Cape Grey on what continent?
   Antarctica

34. The Bingham Canyon Mine, the world's largest manmade excavation is in the Oquirrh Mountains southwest of what major city located on a lake?
   Salt Lake City

35. Khardung La, one of the world's highest navigable mountain passes, is located in what region of the Indian state of Jammu and Kashmir?
   Ladakh

36. The world's largest glacier is located in East Antarctica southeast of the Prince Charles Mountains. Name this glacier.
   Lambert Glacier

37. The Mponeng Gold Mine is the world's deepest mine and is located in what South African province?
Gauteng Province

38. The southernmost point in mainland Asia is at Tanjung Piai, a cape on what peninsula?
Malay Peninsula

39. The world's lowest lake on any island is a lake belonging to the Dominican Republic, on the island of Hispaniola. Name this lake.
Lago Enriquillo

40. What island is the largest in the Caribbean Sea, ahead of Hispaniola?
Cuba

41. Kaffeklubben Island, also known as Coffee Club Island, is home to the northernmost point of land on Earth and belongs to what Danish territory?
Greenland

42. Cape Flissingsky is located on Northern Island in the Novaya Zemlya Archipelago and is the easternmost point on what continent?
Europe

43. Cape Fligely is on Rudolf Island and is the northernmost point in Russia on what continent?
Europe

44. The southernmost point in mainland North America is a cape on the Azuero Peninsula in Panama. Name this cape.
Punta Mariato (Mariato Point)

45. Name the largest saltwater lake in North America and in the Western Hemisphere that is also the largest remnant of the prehistoric Lake Bonneville.
Great Salt Lake

46. The southernmost area of open ocean is at what Antarctic bay that is part of the Ross Sea and is north of Roosevelt Island?
Bay of Whales

47. The lowest point ever reached underground is at the Kola Superdeep Borehole in what country?
Russia

48. The world's highest permanent settlement is at the town of La Rinconada, located near a gold mine in the Andes Mountains of what country?
Peru

49. Cape Dezhnev, the easternmost mainland point in Asia, is on what peninsula in Russia?
Chukchi Peninsula

50. The world's largest lake on an island is a freshwater lake on Baffin Island in the Great Plain of the Koukdjuak. Name this lake.
Nettilling Lake

51. Leadville, the highest city in North America is a settlement near the Arkansas River in what U.S. state?
Colorado

52. Cape Chelyuskin, the northernmost point in mainland Asia, is on what Russian peninsula?
Taymyr Peninsula

53. The Mana Pass, reputed to be the highest vehicle-accessible pass in the world, is located in Nanda Devi Biosphere Reserve, on the border between China and what other country?
India

54. The highest volcano in North America is on the border between the Mexican states of Veracruz and Puebla. Name this volcano, the highest point in Mexico.
Pico de Orizaba (Citlaltepetl)

55. Pamana Island, the southernmost point in Asia, belongs to what archipelagic country?
Indonesia

56. The southernmost point in Europe is an island south of Crete, belonging to Greece. Name this island.

Gavdos Island

57. Name the largest island in a lake in an island in a lake in the world.
Manitou Lake

Geographic Estimation

In this round, your goal is to try to get as close to the correct number as possible. These questions are among the most difficult at the National Geographic Bee and NSF Senior GeoBee Phase 3 Round.

1. What is the height of Mount Everest, in feet?
29,029 feet

2. What is the diameter of the Earth, in miles?
7,917.5 miles

3. What is the depth of Challenger Deep in the Mariana Trench, in feet?
36,070 feet

4. Rounding to the nearest thousand, what is the approximate length of the Mid-Atlantic Ridge, in miles?
10,000 miles

5. What is the area of the continent of Antarctica, in sq mi?
5.405 million sq mi or 5,405,000 sq mi

6. What is the height of Mount Kilimanjaro, in feet?
   19,341 feet

7. How many lakes are there in Finland?
   187,888

8. How many mountain peaks stand taller than 8,000 meters, giving them the name "Eight-Thousanders"?
   Fourteen

9. Given by the World Factbook, what is the length of Canada's coastline, in kilometers?
   202,080 kilometers

10. What is the length of the Nile River, in miles?
    4,258 miles

11. What area of South America does the Amazon River drain, in sq mi?
    2,722,000 sq mi

2013 National

Geographic Bee Finals

Competition

1. What country is bordered by Burkina Faso and Libya?
   Niger

2. What country is bordered by Thailand and Cambodia?
   Laos

3. What country is bordered by Turkey and Saudi Arabia?
   Iraq

4. What country is bordered by Panama and Nicaragua?
   Costa Rica

5. What country is bordered by Somalia and Kenya?
   Ethiopia

6. What country is bordered by North Korea and Myanmar?
   China

7. What country is bordered by Angola and Malawi?
   Zambia

8. What country is bordered by Mexico and Guatemala?
   Belize

9. What country is bordered by Belarus and Romania?
   Ukraine

10. What country is bordered by Ecuador and Peru?
    Colombia

11. What country is bordered by Lithuania and Germany?
    Poland

12. Mountaineer Gerlinde Kaltenbrunner became the first woman to climb the world's 14 highest peaks without using supplemental oxygen. The final peak in her expedition was K2, located on the border between China and what other country?
    Pakistan

13. K2 lies in what mountain range that is an extension of the Hindu Kush mountain system?
    Karakoram Range

14. Geneticist Spencer Wells explores the human past by collecting DNA samples from around the world. One group studied is a segment of the Bushman population

that lives in the desert region west of the Caprivi Strip in what country?
Namibia

15. Northern Namibia is the site of one of the world's largest wildlife parks, which is centered on a large salt pan. Name this salt pan.
Etosha Pan

16. Botanist Joseph Rock, posing here in the 1920's with the king of Muli, made many expeditions to the upper Salween River region near the Tanggula Range. This range lies in what present day country?
China

17. Joseph Rock studied plant life throughout southwest China, including what province that borders Myanmar, Laos, and Vietnam?
Yunnan Province

18. Conservationists Beverly and Derek Joubert are raising awareness about the decline of big cats in the wild. These explorers are studying lions in the Okavango Delta in what country?
Botswana

19. The Okavango Delta feeds into what lake to the south?
Lake Ngami

20. Adventurer Kira Salak, pictured here with her guide, walked in the footsteps of 19th century explorer Hugh Clapperton through the historic Tripolitania region of what country?
Libya

21. Name the historic region south of Tripolitania that includes the Saharan oases of Sabha and Marzuq.
Fezzan

22. Biologist Roman Dial climbs a mountain eucalyptus tree while performing an ecological survey in the Hume Plateau, located in the Great Dividing Range in what county?
Australia

23. Part of Dial's research took place near the Australian Alps, which stretch from New South Wales into what other state?
Victoria

24. In 1938, archaeologist Matthew Stirling uncovered giant stone heads from the Olmec people in the village of La Venta, located on the Isthmus of Tehuantepec in what country?
Mexico

25. These stone heads were moved to museums in the capital of the state of Tabasco. Name this city.
Villahermosa

26. Conservationist Mike Fay displays his field journal while documenting one of the world's largest remaining concentrations of elephants in Zakouma National Park. This park is located southeast of the city of N'Djamena in what country?
Chad

27. The city of N'Djamena is located at the confluence of the Logone River and what other river?
Chari River

28. Climber Bradford Washburn mapped and photographed mountains throughout the world, including Mount Logan, the highest peak in what Western Hemisphere country?
Canada

29. Situated in Yukon's southwest corner, Mount Logan is located in what subrange of the Coast Ranges?
St. Elias Mountains

30. In the late 1800's, Norwegian explorer Fridtjof Nansen led an expedition to the North Pole, reaching a record northern latitude. On his retreat, Nansen crossed Franz Josef Land, an archipelago that belongs to what present day country?
Russia

31. Nansen was rescued from Franz Josef Land and returned to the city of Vardo in northeastern Norway. What sea lies between Franz Josef Land and Vardo?
Barents Sea

32. Conservation photographer Steve Winter has documented the tension between wildlife and humans in Gunung Leuser National Park, located northwest of the Barisan Mountains in what island country?
Indonesia

33. The Barisan Mountains are located on what island?
Sumatra

34. Boa constrictors are found throughout tropical Central and South America, and are now an invasive species on the westernmost island of the Lesser Antilles. Name this island.
Aruba

35. Diamonds are exported from what city in South Africa?
Port Elizabeth

36. Port Elizabeth is located on what bay?
Algoa Bay

37. Sugar is exported from what city in Indonesia?
Surabaya

38. Surabaya is located at the western end of what strait?

Madura Strait

39. What port city is a busy cruise ship terminal in the United Kingdom?
Southampton

40. Southampton is located north of what island in the English Channel?
Isle of Wight

41. Coffee is exported from what city in Ecuador?
Guayaquil

42. What island lies opposite of the mouth of the Guayas River at the head of the Gulf of Guayaquil?
Puna Island

43. Automobiles are exported from what city in South Korea?
Busan

44. Busan lies at the mouth of South Korea's longest river. Name this river.
Nakdong River

45. Glazed tiles are exported from what city in Spain?
Valencia

46. Valencia is located near the mouth of what river in Spain?
Turia River

47. Refined petroleum is exported from what city in Egypt?
Alexandria

48. Alexandria is northeast of a basin that is located in the Libya Desert. Name this basin.
Qattara Depression

49. Electronics are exported from what city in China?
Qingdao

50. Qingdao lies on the southern coast of what peninsula?
Shandong Peninsula

51. Cacao is exported from what city in Brazil?
Salvador

52. Salvador is located on what bay in Brazil?
Todos os Santos Bay

53. Natural gas is shipped from what port city in Australia?
Darwin

54. Darwin lies just south of the Tiwi Islands. Name the largest of these islands.
Melville Island

55. Textiles are exported from what port city in Iran?
Bandar 'Abbas

56. Bandar Abbas lies north of the largest island in the Strait of Hormuz. Name this island.
Qeshm

57. The whale shark can be found seasonally in the Mesoamerican reef system near Isla Holbox off the coast of what peninsula?
Yucatan Peninsula

58. The city of Porto lies on what river that flows west into the Atlantic Ocean?
Douro River

59. The Charles Darwin Research Station is located near Puerto Ayora on what island that is the second largest in the Galapagos?
Santa Cruz

60. Machu Picchu overlooks what river that flows into the Ucayali River?
Urubamba River

61. Costa Rica's Corcovado National Park is located on what peninsula that borders Dulce Gulf?
Osa Peninsula

62. The Lemaire Channel lies northeast of what sea that borders Ellsworth Land and Alexander Island?
Bellingshausen Sea

63. The Pearl Islands, with a storied past involving Spanish conquistadors and buccaneers, are located in what large gulf?
Gulf of Panama

64. Shackleton is buried on what island that is administered by the United Kingdom and lies at the northeast corner of the Scotia Sea?
South Georgia

65. Gray whales migrate to Magdalena Bay, located off the coast of what Mexican state that includes the city of La Paz?
Baja California Sur

66. Name the capital of the British Overseas Territory that is located off the coast of Argentina?
Stanley

67. The oldest form of Tai Chi traces its roots back to a village near the city of Zhengzhou, located near the Yellow River in what province?
Henan Province

68. According to legend, a Taoist priest named Zhang Sanfeng is credited as the creator of Tai Chi. Some believe that he developed many of the movements of Tai Chi at a monastery in the Wudang Mountains, located just south of what river that is a tributary of the Yangtze?
Han River

69. Place these countries in order according to their land area, from largest to smallest: Iran, Yemen, Egypt.
Iran, Egypt, Yemen

70. Which of those countries - Iran, Egypt, Yemen - has the highest population density?
Egypt

71. Place these major cities in order according to their longitude, from west to east: Libreville, Lagos, Bangui.
Lagos, Libreville, Bangui

72. Place these lakes in order according to their surface area from largest to smallest: Nicaragua, Kyoga, Balkhash.
Balkhash, Nicaragua, Kyoga

73. Which of those lakes – Nicaragua, Kyoga, Balkhash - lies closest to the Equator?
Lake Kyoga

74. Place these major cities in order according to their rainfall, from most to least: Dublin, Tokyo, Sofia.
Tokyo, Dublin, Sofia

75. Which of those cities – Dublin, Tokyo, Sofia - has a marine west coast climate?
Dublin

76. Place these major rivers in order according to their length, from longest to shortest: Magdalena, Indus, Yellow.
Yellow, Indus, Magdalena

77. Which of those rivers – Magdalena, Indus, Yellow - has its mouth located farthest south?
Magdalena River

78. Place these major cities in order according to their latitude, from north to south: Budapest, Prague, Zagreb.
Prague, Budapest, Zagreb

79. Which of those cities - Budapest, Prague, Zagreb - has the largest population?
Budapest

80. Place these countries in order according to their GDP per capita: Argentina, Denmark, Slovenia.
Denmark, Slovenia, Argentina

81. Which of those countries has a constitutional monarchy as its form of government?
Denmark

82. Place these islands in order according to their land area, from largest to smallest: Cyprus, Halmahera, Taiwan.
Taiwan, Halmahera, Cyprus

83. Which of those islands - Cyprus, Halmahera, Taiwan - is crossed by the Tropic of Cancer?
Taiwan

84. Place these countries in order according to their population density, from the most densely populated to the least: Belgium, Slovakia, Italy.
Belgium, Italy, Slovakia

85. Which of those countries has the highest percent urban population?
Belgium

86. African penguins live in the waters around southern Africa, including what island that lies at the mouth of Table Bay?
Robben Island

87. Another colony of African penguins can be found on the mainland near Simon's Town, a seaside resort located on what bay?
False Bay

88. Which city is the odd one out? Alicante, Coimbra, Malaga, Valladoid.
Coimbra - the other cities are in Spain, while Coimbra is in Portugal

89. Which sect is the odd one out? Shaktism, Shia, Sufism, Sunni.

Shaktism - the other branches are branches of Islam, while Shaktism is a branch of Hinduism

90. Which city is the odd one out? – Bengkulu, Bandung, Medan, Banda Aceh.
Bandung - the other cities are located on Sumatra, while **Bandung is located on Java**

91. Which city is the odd one out? – Barranquilla, Buenaventura, Cartagena, Cienaga.
Buenaventura - the other cities are located on the Caribbean Sea, while Buenaventura is located on the Pacific Ocean

92. Name the Odd Item Out: Altay Mountains, Rhodope Mountains, Kopetdag Mountains, Tian Shan.
Rhodope Mountains - the other mountain ranges are located in Asia, while the Rhodope Mountains are located in Europe

93. Name the Odd Item Out: Lake Van, Lake Tuz, Lake Urmia, Lake Iznik.
Lake Urmia - the other lakes are located in Turkey, while Lake Urmia is located in Iran

94. Name the Odd Item Out: Sabarmati River, Krishna River, Narmada River, Tapi River.
Krishna River - the other rivers flow into the Arabian Sea, while the Krishna River flows into the Bay of Bengal

95. What river is a heavily trafficked waterway, was once controlled by the Ancient Gauls, and rises in the Langres Plateau to then flow about 480 miles?
Seine River

96. The Pont de Normandie, one of the longest cable-stayed bridges in the world, spans the mouth of what river?
Seine River

97. The Cathedral of Notre Dame is located on an island in what river rising in the Langres Plateau?
Seine River

98. Located in the Simien Mountains, Ras Dejen is the highest peak in what country?
Ethiopia

99. One of the world's largest deposits of rare earth elements, which are used in the production of many high tech gadgets, is located near the largest city in China's Inner Mongolia Autonomous Region. Name this city.
Baotou

100. A capital city on the Arabian Peninsula located at about 7,200 feet receives its water supply from an aquifer that is forecast to run dry in the next decade. Name this capital city.
Sanaa

101. Name the oil-rich exclave that lies just north of the mouth of the Congo River.
Cabinda

102. Because Earth bulges at the Equator, the point that is farthest from Earth's center is the summit of a peak in Ecuador. Name this peak.
Chimborazo

2014 National Geographic Bee Finals Competition

1. Name the deepest lake in the world.
    Lake Baikal

2. The driest place in the world is in which South American desert?
    Atacama Desert

3. Name the highest mountain in Antarctica.
    Vinson Massif

4. Name the world's most densely populated country.
    Monaco

5. Name the longest river in Asia.
    Yangtze River

6. Name the world's highest uninterrupted waterfall.
    Angel Falls

7. Name the largest island within the Indian Ocean.
Madagascar

8. The lowest point on the continent of Australia is what lake?
Lake Eyre

9. Name the world's smallest country in area.
Vatican City

10. Name the most populous city in South America.
Sao Paulo

11. Photographer Nick Nichols used a drone camera to get up close and personal with lions near the Seronera River in what national park located west of Lake Natron?
Serengeti National Park

12. Last summer (2013), astronauts captured images of flooding in a Canadian city located at the confluence of the Bow and Elbow Rivers. Name this city.
Calgary

13. What city includes a mosque built on top of the ruins of an ancient temple, with hundreds of sphinxes stretching into the distance?
Luxor

14. Caboolture is located in the Moreton Bay Region just north of which Australian state capital city?
Brisbane

15. Paella, a traditional rice dish, originated near the Albufera lagoon, just south of what city at the mouth of the Turia River?
Valencia

16. Kaiseki, a multi-course meal with an artistic presentation, is popular in what former capital city located on the Kamo River?
Kyoto

17. Kaiseki often includes sushi and what other Japanese delicacy that consists of raw fish sliced into thin pieces, but served without rice?
Sashimi

18. Crayfish parties, called Kraftskiva, are popular throughout Scandinavia in late summer, including what city located on the easternmost bay of Lake Malaren?
Stockholm

19. Sate Pedang is one variation of skewered meat served with sauce in the most populous city on the island of Sumatra. Name this city.
Medan

20. Marzipan, a sugary treat painted to look like fruit, is popular in what city located near Mount Pellegrino on the island of Sicily?
Palermo

21. Chiles en Nogada, a dish of meat-filled chili peppers in walnut cream sauce, represents the colors of the Mexican flag and is popular in the country's fourth largest city. Name this state capital city located east of Popocatepetl.
Puebla

22. Couscous, a traditional Berber dish made from wheat, is sometimes served in what African capital city located near Cape Bon?
Tunis

23. Earth's diameter at the Equator is slightly larger than its diameter at the Poles. To the nearest whole number, give the equatorial diameter of the Earth in miles.
7926 miles

24. This island, home to Southwest National Park, was a British colony during most of the 19th century. Its highest point rises to about 5,300 feet. The island was formerly known as Van Diemen's Land.
Tasmania

25. Mawsynram, India, has a lengthy monsoon season and is often referred to as the wettest place on Earth. On

average, how many inches of rainfall does Mawsynram receive annually?

467 inches

26. Last summer (2013), Crown Prince Philippe, the Duke of Brabant, became king of a European country after his father abdicated for health reasons. Name this country.

    Belgium

27. Oyala, a planned city located in the rainforest 65 miles east of Bata, is being built as a new capital for which African country?

    Equatorial Guinea

28. The discovery of a major shale oil deposit in the Vaca Muerta formation in 2010 has led to an expansion of oil drilling in the Neuquen Province in what country?

    Argentina

2015 National

Geographic Bee Finals

Competition

1. Cadillac Ranch, which features ten Cadillacs covered in spray paint graffiti, is located near the largest city in the Texas panhandle. Name this city.
   Amarillo

2. This dinosaur sculpture attracts tourists to South Dakota's second largest city. Name this city.
   Rapid City

3. A 31-foot-tall statue of Paul Bunyan is located in Maine in the largest city on the Penobscot River. Name this city.
   Bangor

4. The world's largest rubber stamp is located in Ohio in the largest city on the Cuyahoga River. Name this city.
   Cleveland

5. Eight-foot-tall painted cowboy boots can be found throughout the largest city in Wyoming. Name this city.
    Cheyenne

6. Cupid's Span, a 60-foot-tall bow and arrow sculpture, is located near Fisherman's Wharf and the Presidio in one of California's largest cities. Name this city.
    San Francisco

7. A 76-foot-tall statue of an oil driller stands at Expo Square in Oklahoma's second largest city. Name this city.
    Tulsa

8. A 40-foot-tall blue bear peers through the windows of a convention center in the largest city in the Front Range. Name this city.
    Denver

9. An 800-pound statue of a dog stands guard at a trailer company in the largest port city in Georgia. Name this city.
    Savannah

10. This giant spoon and cherry decorate a sculpture garden in Minnesota's largest city. Name this city.
    Minneapolis

11. Wynton Marsalis performed at a festival in a city on Narragansett Bay, located on the southern end of Aquidneck Island. Name this city.

Newport

12. Name the Indonesian mountain range that stretches the length of the island of Sumatra where Titan arum can be found.
Barisan Mountains

13. Name the country where two European scientists invented flashlight powder in 1887 near the city of Potsdam.
Germany

14. Name the river along which the Havasupai tribe lives on the Coconino Plateau.
Colorado River

15. Plastic marine debris is threatening wildlife on what small group of U.S. – administered islands south of Kure Atoll?
Midway Islands

16. What U.S. city is located at the mouth of the Patapsco River?
Baltimore

17. Name the highest mountain in Oregon.
Mount Hood

18. Louisiana's flag depicts what animal?
Pelican

19. Name the largest city in the San Joaquin Valley.
Fresno

20. Wilmington, North Carolina, is located on what river?
Cape Fear River

21. What crop is the most widely planted field crop in the United States?
Corn

22. What U.S. city is located at the head of Cook Inlet?
Anchorage

23. The Pecos River is a tributary of what other river?
Rio Grande

24. What is the term for a narrow strip of land that connects two larger landmasses?
Isthmus

25. What Iowa City is located across the Missouri River from Omaha?
Council Bluffs

26. The lowest point in Idaho is on what river?
Snake River

27. The Mesabi Range contains a large deposit of what metal-bearing mineral?

Taconite

28. What city is located near the confluence of the McKenzie and Willamette Rivers?
    Eugene

29. The Keweenaw Peninsula juts into what large lake?
    Lake Superior

30. How many U.S. states have populations over 15 million?
    Four

31. Name the largest city on the Big Island of Hawaii.
    Hilo

32. What mountain is the highest point in Maine?
    Mount Katahdin

33. What is the term for a line on a map that connects points of equal temperature?
    Isotherm

34. What river connects Lake St. Clair to Lake Erie?
    Detroit River

35. The Seward Peninsula borders what strait?
    Bering Strait

36. What legume is the official state crop of Georgia?
    Peanut

37. What U.S. city is located where the Fox River enters Lake Winnebago?
   Oshkosh

38. What lake is the lowest point in Vermont?
   Lake Champlain

39. The Powder River Basin is best known for its large deposits of what fossil fuel?
   Coal

40. Name the largest city on Humboldt Bay.
   Eureka

41. The Niobrara River is a tributary of what larger river?
   Missouri River

42. How many U.S. states border the Gulf of Mexico?
   Five

43. What major city is located near the mouth of the Genesee River?
   Rochester

44. What dam created Lake Mead?
   Hoover Dam

45. South Carolina's flag depicts what tree?
   Palmetto Tree

46. Australia's Geographe Bay, a popular vacation spot, is located about 150 miles south of which state capital city?
Perth

47. Italy's chief port city is on a gulf of the Ligurian Sea. Name this port city.
Genoa

48. A desert, which has a name meaning "Black Sands," covers more than 70 percent of Turkmenistan. Name this desert.
Karakum Desert

49. Fishing and sheep raising are the chief economic activities in a group of Danish islands that lies northwest of the Shetland Islands. Name these islands.
Faroe Islands

50. Lake Gatun, an artificial lake that constitutes part of the Panama Canal system, was created by damming which river?
Chagres

51. Which city on the Lena River is an important fur-trading center in Siberia?
Yakutsk

52. The returned sarcophagus will be housed in the Grand Egyptian Museum on the outskirts of what city located 3 miles southwest of Cairo?
Giza

53. The painting 'A Sunday Afternoon on the Island of La Grande Jatte' depicts a scene on an island located in what river?
Seine River

54. French fries may have originated along the Meuse River near the city of Liege in which country?
Belgium

55. What is the most densely populated country in Central America?
El Salvador

56. What is the capital of the Marshall Islands?
Majuro

57. What is the official language of St. Lucia?
English

58. Lake Ohrid straddles the border between Albania and what neighboring country?
Macedonia

59. Peru's chief seaport is just west of Lima. Name this city.
Callao

60. What is the official religion of Bangladesh?
    Islam

61. The atoll of Funafuti is part of which country?
    Tuvalu

62. What river flows through the city of Astrakhan?
    Volga River

63. What is the official currency of Liechtenstein?
    Swiss franc

64. Tripolitania is a historic region in which country?
    Libya

65. Name the second largest city on the island of Tasmania.
    Launceston

66. What is the official religion of Sri Lanka?
    Buddhism

67. The island of Gozo is part of which Mediterranean country?
    Malta

68. What is the capital of Timor-Leste?
    Dili

69. What is the currency of Mongolia?

Mongolian togrog

70. A causeway connects Saudi Arabia and what country?
Bahrain

71. Name the only national capital city on the island of New Guinea.
Port Moresby

72. What is the official language of Togo?
French

73. Mariupol, a city located at the mouth of the Kalmius River, is located on what sea that is an arm of the Black Sea?
Sea of Azov

74. Helsingor's strategic located on a narrow strait allowed Danish kings to collect tolls from passing ships. Name this strait.
Oresund

75. A Russian island that straddles 180 degree longitude is one of the most biodiverse in the Arctic and is the world's northernmost UNESCO World Heritage Site. Name this island.
Wrangel Island

76. In 2014, the government of India established a new state out of the northwestern part of Andhra Pradesh. Name this new state.
Telangana

77. The Strait of Canso separates mainland Canada from what island?
Cape Breton Island

78. What Central Asian capital city is located northwest of the densely populated Fergana Valley?
Tashkent

79. If completed, the proposed Grand Inga Dam would become the world's largest hydropower plant. This dam would be built near Inga Falls on which African River?
Congo River

2016 National Geographic Bee Finals Competition

1. Windsurfing conditions are ideal on the Columbia River near which Washington city located just north of Portland?
    Vancouver

2. You can go ice climbing on frozen waterfalls near which Rocky Mountain city that lies 12 miles east of Pikes Peak?
    Colorado Springs

3. Hiking is a popular activity in a city known as the American Riviera, located between the Pacific Ocean and the Santa Ynez Mountains about 80 miles northwest of Los Angeles. Name this city.
    Santa Barbara

4. Canoeing along Barton Creek offers scenic views of the largest city on the Colorado River, which is the longest river entirely in Texas. Name this city.

Austin

5. Several nearby mountain ranges offer countless trails for mountain biking near Montana's second largest city. Name this city.
    Missoula

6. Sea kayakers can view spectacular glaciers near the largest port city on Prince William Sound in Alaska. Name this city.
    Valdez

7. Rappelling is just one of many skills required when canyoneering in eastern Utah near which city located just south of Arches National Park?
    Moab

8. Cave divers can explore underwater rock formations near the largest city in northwest Florida. Name this city located north of Apalachee Bay.
    Tallahassee Bay

9. You can snowboard in the San Francisco Peaks, a mountain range in Arizona just north of which city situated at the base of Mount Elden?
    Flagstaff

10. Some of the best whitewater rafting in the country can be found in North Carolina near a city 19 miles southwest of Mount Mitchell. Name this city.

Asheville

11. Many sea turtles nest on what cape in Everglades National Park that is the southernmost extremity of the U.S. mainland?
Cape Sable

12. Bronze helmets dating back to the 6th century B.C. were discovered in a cemetery near Pella, the capital of what ancient kingdom once ruled by Alexander the Great?
Macedon

13. The drop-cam was recently used during an expedition to the Revillagigedo Islands. These islands are located about 250 miles south of what large peninsula?
Baja California

14. Giant otters can be found in the Manu River and what other river that shares its name with the Peruvian region where most of Manu National Park is located?
Madre de Dios River

15. Nine-banded armadillos can be found on what U.S. plateau that is the source of the Llano River?
Edwards Plateau

16. Name the highest mountain in Washington.
Mount Rainier

17. Name the most populous city on the Cumberland River.

Nashville

18. Wyoming's flag depicts what animal?
American Bison

19. Lake Placid and Saranac Lake are located in which mountain range?
Adirondack Mountains

20. Shoshone Falls is a waterfall located on what river?
Snake River

21. Wisconsin leads the nation in the production of what fruit?
Wisconsin

22. The city of Nome is located on which Alaskan peninsula?
Seward Peninsula

23. The Connecticut River empties into what sound?
Long Island Sound

24. What is the term for a body of water that is enclosed by an atoll?
Lagoon

25. San Clemente and San Nicolas are part of what island chain?
Channel Islands

26. What large city is located at the confluence of the Ohio and Little Miami Rivers?
Cincinnati

27. Name the state gem of Arkansas, which is featured on its flag and state quarter.
Diamond

28. The Apostle Islands are located in which Great Lake?
Lake Superior

29. What city is located on a peninsula between the Cooper and Ashley Rivers?
Charleston

30. How many U.S. state capitals are located on the Mississippi River?
Two

31. Name the largest city on the island of Puerto Rico.
San Juan

32. The Roanoke River empties into what sound?
Albemarle Sound

33. What is the term for a bowl-shaped depression made by the impact of a meteorite on Earth's surface?
Crater

34. The Cape Cod Canal connects Cape Cod Bay with which other bay to the southwest?
Buzzards Bay

35. Shasta Dam in California is located on what river?
Sacramento River

36. What animal is the official state crustacean of Maryland?
Crab

37. Michigan's lowest point is at which Great Lake?
Lake Erie

38. What major U.S. city is located at the confluence of the Wolf and Mississippi Rivers?
Memphis

39. Alaska is the leading producer of what metallic element that has the symbol Ag?
Silver

40. What mountain range is located just east of Great Salt Lake?
Wasatch Range

41. The city of Richmond, Virginia is located on what river?
James River

42. How many U.S. states border Lake Michigan?
Four

43. Name the highest mountain in the state of Hawaii.
    Mauna Kea

44. What Texas city is located near the mouth of the Nueces River?
    Corpus Christi

45. The reverse side of the Oregon state flag features which mammal?
    Beaver

46. According to the U.S. Geological Survey, how many earthquakes of magnitude 6.0 or above were recorded worldwide in 2015?
    146

47. You can paraglide in front of Jungfrau Mountain, located south of the Swiss resort town of Interlaken in which mountain range that is part of the Alps?
    Bernese Alps

48. Slacklining, similar to tightrope walking, is a popular activity over the Andaman Sea near Thailand's largest island. Name this island.
    Phuket Island

49. You can try out dog sledding on the frozen lakes near the largest city in Canada's Northwest Territories. Name this city.

Yellowknife

50. You can get up close and personal with great white sharks while cage diving near Danger Point, located near what cape that is the southernmost point of Africa?
Cape Agulhas

51. Catch views of Waitemata Harbour while jumping off of Sky Tower, located in which city that lies on an isthmus of North Island?
Auckland

52. Flyboarding, a sport in which a water jetpack propels you into the air on a hoverboard, is gaining popularity in Europe, including areas around which city near the mouth of the Douro River?
Porto

53. Friesian horses originated in Friesland, which borders the largest lake in the Netherlands. Name this shallow manmade lake.
Ijsselmeer

54. Name this Mediterranean port city located on the northern slopes of Lake Carmel.
Haifa

55. Name the tributary that rises in central Angola and joins the Okavango River – also known as the Cubango River – at Angola's southern border.

Cuito River

56. Dirk Hartog Island is located off the coast of what country?
Australia

57. What is the capital of St. Kitts and Nevis?
Basse-Terre

58. What is the official language of Suriname?
Dutch

59. What is the most densely populated country in South America?
Ecuador

60. One of Mexico's busiest seaports is located in the state of Colima. Name this city.
Manzanillo

61. What is the official religion of Cambodia?
Buddhism

62. Prince Patrick Island is part of which country?
Canada

63. What river flows through the city of Dresden?
Elbe River

64. What is the official currency of Algeria?

Algerian dinar

65. Ludogorie is a region in which European country?
Bulgaria

66. Name the largest city on the island of Sulawesi.
Makassar

67. What is the official religion of Mauritania?
Islam

68. The Pelagian Islands belong to which country?
Italy

69. What is the capital of St. Lucia?
Castries

70. What is the official language of Sierra Leone?
English

71. Name the largest city on Kola Fjord, an inlet of the Barents Sea.
Murmansk

72. Name the national capital city located on the island of Bioko.
Malabo

73. What is the currency of Poland?
Polish zloty

74. The Gotthard Base Tunnel, expected to open in early June, will be the world's longest rail tunnel. This tunnel is located in which country?
Switzerland

75. The Asian island of Bangka, a major tin producing center, is just off the eastern coast of what larger island?
Sumatra

76. Name the dry and dusty trade wind that originates in the Sahara and blows over West Africa, often causing airport delays in winter.
Harmattan

77. The ruins of Ephesus, important in the history of Christianity, are located in which present day country?
Turkey

78. An active lighthouse is located on a cape that is the easternmost point of mainland Australia. Name this cape.
Cape Byron

79. A new marine sanctuary will protect sharks and other wildlife around Isla Wolf in which archipelago in the Pacific Ocean?
Galapagos Islands

80. Which East African lake that drains into the Ruzizi River contains large quantities of dissolved methane gas that could generate electricity for millions of people?
Lake Kivu

2017 National Geographic Bee Finals Competition

1. More than 1,000 species of plants thrive in the bogs and estuaries on the largest island off the coast of Maine. Name this island.
   Mount Desert Island

2. Trempealau, Chickasaw, and Yazoo are all wildlife refuges for birds migrating along the course of what river?
   Mississippi River

3. Alaska's Kootznoowoo Wilderness area, located on Admiralty Island, has one of the highest concentrations of brown bears in the world. This wilderness area lies 50 miles south of what important city?
   Juneau

4. In the shadow of Mount Hood, the waterfalls of Mark O. Hatfield Wilderness Area cascade over basalt cliffs before draining into what river?
    Columbia River

5. More than 40 species of cacti can be found in the Guadalupe Mountains, located in a desert that extends from Mexico into Texas, New Mexico, and Arizona. Name this desert.
    Chihuahuan Desert

6. The glacial forces that created the waterways and rocky outcrops that define the Boundary Waters Canoe Area Wilderness also created a large lake near its eastern border. Name this lake.
    Lake Superior

7. Located about 200 miles west of the Missouri River, what major city serves as the gateway to the Black Hills and to one of the largest cave systems in the world?
    Rapid City

8. The Flint Hills, home to some of the last tallgrass prairie in North America, stretch from northern Kansas south to what river that runs through Wichita?
    Arkansas River

9. The Balcones Canyonlands National Wildlife Refuge, an important nesting site for birds, is located near what river that flows through Texas into Matagorda Bay?

Colorado River

10. St. Marks Wilderness Area, located in the Florida panhandle, protects about 40 miles of coastline along what bay that receives the St. Marks River?
Apalachee River

11. The expedition set off from what small part of Cape Cod that shares its name with a nearby oceanographic institution?
Woods Hole

12. Buck Island and Buck Island Reef National Monument are located just off the coast of what Caribbean Island?
St. Croix

13. The Rio Grande separates Ciudad Juarez from what city?
El Paso

14. The Truckee River drains what large lake?
Lake Tahoe

15. Mississippi's state nickname mentions what flowering tree?
Magnolia

16. What port is the terminus for the Trans Alaska Pipeline System that starts in Prudhoe Bay?
Valdez

17. Wheeling, West Virginia, is located on what river?
   Ohio River

18. What is the official animal of Maine?
   Moose

19. Name Alaska's largest island.
   Kodiak Island

20. What is the highest point in West Virginia?
   Spruce Knob

21. What landform is a tall, steep-sided tower that eroded from a mesa?
   Butte

22. Name the most populous city in Connecticut.
   Bridgeport

23. What narrow body of water separates the Upper and Lower Peninsulas of Michigan?
   Straits of Mackinac

24. The New Jersey state flag features the head of which animal?
   Horse

25. The city of Fredericksburg, Virginia, is located on what river?
   Rappahannock River

26. Mount Rainier National Park is located in what mountain range?
Cascade Range

27. Anthracite and lignite are varieties of which fossil fuel?
Coal

28. The Cape May Peninsula borders what bay?
Delaware Bay

29. Name the most populous island in the Northern Mariana Islands.
Saipan

30. What term is given to a continuous line of thunderstorms advancing ahead of a cold front?
Squall Line

31. Beaver Island is the largest island in which of the Great Lakes?
Lake Michigan

32. What state capital is located on the Susquehanna River?
Harrisburg

33. What insect structure is featured on Utah's state flag?
Beehive

34. Name the largest city on Lake Erie.

Cleveland

35. Fort Peck Dam is located on what river?
Missouri River

36. Name the blue-to-green mineral that is New Mexico's state gem.
Turquoise

37. Name the highest mountain in New Hampshire.
Mount Washington

38. Located in Utah, Mount Nebo is the highest peak in which mountain range?
Wasatch Range

39. Name the state tree of Vermont, which is featured on its state quarter.
Sugar Maple

40. The city of Grand Marais is located near the Sawtooth Range on what lake?
Lake Superior

41. What Illinois city is located at the confluence of the Ohio and Mississippi Rivers?
Cairo

42. Maine is the leading producer of which large marine crustacean?

Lobster

43. Using the Great Circle Route, how many miles is it from the United States Capitol Building to the Palace of Westminster?
3,662 miles

44. Paul Salopek recently walked through a desert located between the Amu Darya and Syr Darya. Name this desert.
Kyzylkum Desert

45. Croatia's Plitvice Lakes National Park is famous for its natural limestone dams and waterfalls. This park is located in which mountain range that stretches from northeastern Italy to Albania?
Dinaric Alps

46. Namibia's Skeleton Coast, one of the world's largest unspoiled shorelines, is home to fur seals, elephants, and baboons. It extends from the Angolan border south to what bay that is sheltered by Pelican Point?
Walvis Bay

47. Bialowieza Forest, located on the border between Poland and Belarus, is home to one of Europe's last old-growth forests. It's located 40 miles north of what Belarusian city on the Bug River?
Brest

48. Unique rock formations and hot springs can be found near the city of Keelung along the coast of an Asian island north of the Luzon Strait. Name this island.
Taiwan

49. Brazil's Pantanal region is a sprawling wilderness that is home to the elusive jaguar. The gateway to this region is what city that is the capital of the state of Mato Grosso?
Cuiaba

50. Koalas and penguins can be seen on the rugged coastline of Australia's Phillip Island. This island is located north of what strait?
Bass Strait

51. The James Webb Telescope will be launched next year from a space center near Kourou in which overseas department of France?
French Guiana

52. Name the highest point in Argentina.
Mount Aconcagua (Cerro Aconcagua)

53. The Oder River flows through a lagoon into what sea?
Baltic Sea

54. What is the official language of Brunei?
Malaysian

55. The Volga River flows into what body of water?

Caspian Sea

56. Isla de la Juventud is part of what country?
    Cuba

57. What is the official religion of Tunisia?
    Islam

58. Name the capital of the Gambia.
    Banjul

59. What strait separates Sicily from mainland Italy?
    Strait of Messina

60. What is the official currency of Armenia?
    Armenian dram

61. What lake is the source of the Mackenzie River?
    Great Slave Lake

62. What sea lies to the north of the island of New Britain?
    Bismarck Sea

63. What is the dominant religion of Myanmar?
    Buddhism

64. Name the capital of the Comoros.
    Moroni

65. Name the highest point in Turkey.

Mount Ararat

66. What is the official language of Bhutan?
    Dzongkha

67. The historical region of Attica is part of which country?
    Greece

68. What river flows through Madrid, Spain?
    Manzanares River

69. What is the currency of Bangladesh?
    Bangladeshi taka

70. Scientists are planning to reintroduce tigers to Central Asia 50 years after they became extinct in the region. One possible site for reintroduction is the Almaty Region in what country?
    Kazakhstan

71. Located on the Parana River, one of the world's largest hydroelectric dams has a name that means "singing stone" in the Guarani language. Name this dam.
    Itaipu Dam

72. Mu Gia Pass, a strategic pass and a key point of entry to the Ho Chi Minh Trail, lies in what mountain range?
    Annamite Range

73. Tourists often reach Olmec and Mayan ruins in the Mexican state of Tabasco by going through the state's largest city. Name this city on the Grijalva River.
Villahermosa

74. A small island in the Lesser Antilles is divided politically between two countries. Name this island.
Saint Martin

75. What large mountain system that stretches more than 1,200 miles separates the Taklamakan Desert from the Tibetan Plateau?
Kunlun Shan

After the National

Geographic Bee

*Note: This section is for those who are done participating in the National Geographic Bee.

If you have mastered this book and the others (Ultimate Preparation Guide and Competitor's Compendium, maybe U.S. Reference Guide), then you have likely been or are a state winner as of now. You likely have placed in the top ten at the National Geographic Bee, or even the top 3.

If you have placed in the top 3, your Geography Bee journey is over. If you are in eighth grade and this was or is going to be your last year, you are not eligible to compete in the National Geographic Bee next year.

So what happens to all of those hours you spent on geography? What can you use it for now?

After the National Geographic Bee, pursuing competitive geography can lead you to ranking nationally and internationally in the United States Geography Olympiad and International Geography Olympiad. You can compete in the International Geography Bee.

These high school geography competitions focus not only on geography, but also on:

- Economics
- Demography
- Statistics
- International Relations/Political Science
- Culturology/Cultural Studies
- History
- Anthropology
- Geology
- Meteorology
- Climatology
- Linguistics

- Ecology/Environmental Studies
- Civics and Government
- Cartography

If you are planning to take geography seriously throughout high school, you might want to consider majoring in geography in college. There are plenty of jobs out there that require geography, which include economists, diplomats, ambassadors, GIS professionals, civil engineers, urban planners, data analysts, environmental lawyers, oceanographers, geologists, statisticians, cartographers, location analysts, political affairs officers, scientists, and of course, geographers.

In high school, for serious students pursuing geography, I would recommend opting for AP Human Geography, AP Economics (Macro and Micro), AP Comparative Government and Politics, AP United States Government and Politics, AP World History, AP European History, AP Environmental Science, AP Statistics, *and/or* AP Language and Culture (choose between French, Spanish, Chinese, Japanese, German, Italian, or Latin).

If some of these AP classes are not offered at your high school but you want to study those areas, you can either try to initiate an AP class in your school, or you can do a lot of self-study at home and register to take the AP test.

Try to find out whether your school offers electives and/or full year classes in Geography, World Studies, Cultural Studies, Economics, Geology, Meteorology, Model UN, International Relations, **and/or** Sociology.

Some of the best undergraduate and graduate programs in geography can be found at Boston University, University of Colorado-Boulder, and University of Maryland-College Park. Central Connecticut State University claims to be home to the oldest undergraduate geography program in New England, and Jacksonville University is also a site of many graduates in geography-related fields.

You can sign up, create a team, and participate in the World Geography Bowl competition (for undergraduate students) at http://worldgeobowl.org/ if you are in college.

GeoJokes

Q. Where am I going to eat lunch?
A. At the New Delhi!

Q. Why did the geography student become a scuba diver?
A. His grades were below sea level!

Q. How do you get into Florida?
A. By using the Keys.

Q. What city is the coldest in Europe?
A. Berlin

Q. Why was the map of Finland so elaborate?
A. It was given a lot of Finnish-ing touches!

Q. Where do people trade stocks?
A. Stockholm.

Q. What country is the world's largest manufacturer of ties?
A. Thailand.

Q. What did I do when I got in trouble?
A. Iran!

Q. What did the teacher tell the students on Friday?
A. Stop being so Fiji-ty!

Q. Why wasn't anyone outside during the monsoon rains?
A. They were Indore.

Q. What are Germans' favorite type of burger?
A. Hamburger.

Q. What did the she say when she won the geography bee?
A. No-rway! I can't Belize this!

Q. Why are you Russian to the water fountain?
A. I ate some really spicy Chile.

Q. What do people use to make their hair sleek?
A. Cologne.

Q. What did the customer say to the waiter?
A. I'm Hungary for some Turkey!

Q. Which country is the most enthusiastic?
A. New Zealand.

Q. Where did you get that Bruges from?
A. I fell down the Grand Staircase. *

Q. Where are people elected?
A. The Presidential Range.

Q. Where did the dolphin deposit its money?
A. At the riverbanks.

Q. What does Wisconsin's cheese sell for a lot of?
A. Moo-lah!

Q. Have you Aden anything today?
A. Yes, I had a banana Split earlier. **

Q. What do you call a map guide to Rikers Island? ***
A. A con-tour map!

Q. What are you going to watch on TV?
A. The English Channel.

Q. Where do fish buy their fins?

A. Finland.

Q. What is that smell?
A. Bir-n-ingham!

Q. What city is known for its cranberry production?
A. Canberra.

Q. What is the geologist's favorite thing to play with?
A. Marbles.

Q. What happened after he broke the vase?
A. His Barents grounded him.

Q. Why didn't I ask you the capital of Alaska?
A. Because Juneau already.

Q. What did Delaware?
A. A New Jersey.

Q. What did the sensible boy say to his ignorant brother?
A. I'm more Prague-matic than you.

Q. So what happened when you broke the globe?
A. Iran, because I was Ghana get in trouble.

*Grand Staircase is part of a national monument in southern Utah.
**Split is the second largest city in Croatia.
***Rikers Island in New York is a famous prison and one of the world's largest correctional facilities.

About the Author

Keshav Ramesh is a 14-year old author of twenty books, including *The Geography Bee Ultimate Preparation Guide, A Competitor's Compendium to the Geography Bee, The Quintessential Questionnaire to the Geography Bee,* and *The Geography Bee Comprehensive U.S. Reference Guide.*

These geography bee books have been designed to prepare students for all levels of the National Geographic Bee (NGB), United States Geography Olympiad (USGO), International Geography Bee (IGB), and North South Foundation (NSF) Junior/Senior Geography Bees.

Keshav competed in the CT Geographic Bee for three years and ranked 16th nationally in the 2016 NSF Senior GeoBee National Competition in Tampa, Florida. He has also taken the USGO/IGB NQEs several times. In addition, Keshav has competed in geology-related events at the CT Science Olympiad State competition.

Keshav lives in Connecticut with his family.

Bibliography

"Geography." *ThoughtCo*. About, Inc., n.d. Web. <https://www.thoughtco.com/geography-4133035>.

National Geographic. *National Geographic Atlas of the World, Tenth Edition*. 10th ed. N.p.: National Geographic Society, 2014. Print.

National Geographic. *National Geographic Kids Ultimate Globetrotting World Atlas*. N.p.: National Geographic Society, 2014. Print.

National Geographic. *National Geographic Kids United States Atlas*. N.p.: National Geographic Society, 2012. Print.

National Geographic. *National Geographic Kids World Atlas, Fourth Edition*. 4th ed. N.p.: National Geographic Society, 2013. Print.

National Geographic Kids. *National Geographic Science Encyclopedia: Atom Smashing, Food Chemistry, Animals, Space, and More!* N.p.: National Geographic Society, 2016. Print.

National Geographic Kids. *National Geographic United States Encyclopedia: America's People, Places, and Events*. N.p.: National Geographic Society, 2015. Print.

Wikipedia. Wikimedia Foundation, n.d. Web. <https://www.wikipedia.org/>.

Wojtanik, Andrew. *National Geographic Bee Ultimate Fact Book: Countries A to Z*. N.p.: National Geographic Society, 2012. Print.

World Atlas. N.p., n.d. Web. <http://www.worldatlas.com/>.

"World Geography." *Fact Monster*. Sandbox Networks, Inc., n.d. Web. <https://www.factmonster.com/world/world-geography>.